Antbirds
& Ovenbirds

Number Thirty-one
The Corrie Herring Hooks Series

UNIVERSITY OF TEXAS PRESS, AUSTIN

Antbirds & Ovenbirds

Their Lives and Homes

Alexander F. Skutch
Illustrations by Dana Gardner

To Rosendo M. Fraga

Printed in the United States of America
First edition, 1996

♾The paper used in this publication meets the minimum requirements of American National Standard for Information Sciences—Permanence of Paper for Printed Library Materials, ANSI Z39.48-1984.

Library of Congress Cataloging-in-Publication Data

Skutch, Alexander Frank, 1904–
Antbirds and ovenbirds : their lives and homes / by Alexander F. Skutch ; illustrated by Dana Gardner. — 1st ed.
p. cm.
Includes bibliographical references (p.) and index.
ISBN 0-292-77699-3 (alk. paper; cloth). — ISBN 0-292-77705-1 (alk. paper; paperback)
1. Formicariidae. 2. Furnariidae. I. Title.
QL696.P2455S48 1996
598.8—dc20 95-40982

♻ Printed on recycled paper.

Contents

Illustrations

Drawings

Tables

Foreword

by David Snow

Over forty years ago, when I was assisting R. E. Moreau in editing *The Ibis* (the journal of the British Ornithologists' Union), among the most exciting and unusual of the contributions that we received were two long papers by Alexander Skutch. One was entitled "How the male bird discovers the nestlings," and dealt in fascinating detail with a facet of avian natural history that had been almost completely overlooked by previous observers: how, in the many species in which nest-building and incubation are undertaken by the female alone, the male, who may never even have seen the eggs, learns that they have hatched and begins to feed the young.

The second paper, "The parental stratagems of birds," brought a wealth of firsthand observation to bear on an even more vital aspect of bird biology. Skutch's observations showed, among much else, how a sitting bird may have to choose between two very different strategies when a predator approaches its nest. First, it may slip quietly off the nest and flee very inconspicuously when the predator is still a good distance away, trusting that neither its movements nor the nest will be noticed. The second strategy involves sitting very tight and still until the last moment, "hoping" that it will not be seen; if it is discovered, it will leave suddenly and conspicuously and attempt to distract the predator by "broken wing" or other distraction displays. Other strategies performed when the predator is at an intermediate distance likely prove fatal.

"Skutch's contribution is most remarkable," wrote Moreau, who had himself made pioneering studies of the breeding of tropical African birds and knew something of the patience needed and discomfort suffered in such field studies, let alone the persistence and experience needed even to find the nests of little-known species in the tropical forest. Forty years later, one can only wonder at the continuing flood of new observations that have enriched our knowledge of the life histories of so many tropical American birds—observations that have in many ways made the richest avifauna in the world one of the most intimately known.

The antbirds and ovenbirds, two of the largest American bird families, the former almost exclusively tropical and the latter for the most part so, have until now had no books devoted to them. Regional handbooks give them summary treatment, listing basic facts but hardly bringing the birds to life. A few papers in ornithological journals deal in detail with a limited number of species, but are not easily accessible to most readers. The two families are placed next to one another in standard classifications, and they share some important and obvious anatomical characters, including a complete absence of bright plumage, fairly simple songs and calls, and largely insectivorous diets.

However, in other ways the two families could hardly be more different. Most striking is the difference in their nesting: antbirds build simple nests, mostly of conventional cup shape, while ovenbirds show an extraordinary development and variety of nest architecture unrivaled by any other bird family. As their name implies, many species of antbirds feed in association with army ants, following their raiding columns and taking the insects that the ants disturb. A few species are "professional" ant-followers, almost wholly dependent on this unusual source of sustenance. Ovenbirds are ecologically more adaptable than antbirds, but they do not follow army ants. They forage in a remarkable number of different ways and in a wide variety of habitats, some ranging far beyond the tropics, a few ascending high into the Andes, a few living in semi-deserts, in marshes, or along sea coasts. Most—perhaps all—species of ovenbirds make use of some kind of communal dormitory for roosting at night, which may be specially built for the purpose. All antbirds, so far as we know, roost solitarily in vegetation. With Skutch as a guide, all this variety is brought vividly to life. Reading his text is the next best thing to seeing the birds firsthand. Indeed, in some ways it is better, as his knowledge enables him to make illuminating comparisons and point out contrasts between one species and another.

Each reader will doubtless find his or her favorite passages in a book that is full of gems of observation and description. One of Skutch's great attributes has been his ability not only to make unique observations, but also to describe them in a style full of fluency and grace. Not for him the dry, jargon-ridden writing that all too often one finds in scientific ornithology. The clarity of his writing is outstanding, too, and it serves him well in his descriptions. For example, he lucidly illustrates the way in which a pair of Rufous-breasted Spinetails construct their elaborate, fortified stick-nest, with the floor of the nest chamber lined "with downy leaves, often of a shrubby *Solanum*" (a characteristic detail, as Skutch is an excellent botanist as well as ornithologist). In addition, he describes an entrance tunnel surrounded by a platform of twigs, for which "cast reptile skins, of which the birds never seem to have enough, are diligently sought, stuffed here and there into the walls, and laid on the entrance tunnel, often until they carpet the whole length of the birds' front hall."

The regular incorporation of snakeskins into their nests by this and other ovenbird species, and the whole extraordinary elaboration of their nests, to which they may devote much of their lives and which seems to have no function, leads Skutch to speculate about their mental powers and to suggest that they may enjoy their nest-building activity in a way not altogether different from the way we enjoy our creative activities.

There is plenty here for evolutionists to think about. Incorporation of snakeskins into ovenbirds' nests may have no positive adaptive function: occasionally, a bird's behavior seems actually maladaptive. The most striking case is that of Hudson's Canastero, an ovenbird whose nests among cardoon thistles in the pampas of Argentina are so well hidden that W. H. Hudson, after whom the species was named, searched every day for a whole season without finding one. Yet the nestlings' continual calling makes them such an easy prey for Chimango hawks to find that they feed their own young little else. "This reminds us . . . how capricious natural selection, which over long ages adapts some organisms so finely to their circumstances, may fail to accomplish something so seemingly simple as silencing nestlings' voices responsible for their wholesale destruction."

At this time detailed field studies concentrating on the natural history and the behavior of fascinating but little-known species are out of fashion and attract little funding. They have given way to statistically based, often experimental studies aimed at testing theoretically derived hypotheses or models. Skutch's book is a striking reminder of the continuing value of what some mistakenly dismiss as old-fashioned natural history. Surely both approaches are still needed. Academic, theoretical biologists need facts of the kind that Skutch provides, facts which few have the opportunity or the persistence to discover.

This book, like so many of Skutch's previous books and papers, gives details of a kind that can hardly be found—and certainly not in such abundance—in other scientific ornithological publications. They are the happenings, the raw material of the evolution of life. Skutch describes them and interprets them as a biologist, and as a philosopher does not hesitate to incorporate them into a wider vision of the evolution and advancement of conscious life.

Beautifully complementing Alexander Skutch's word paintings are Dana Gardner's fine, crisp drawings of the birds that are the subject of this book. Together the author and the artist present a valuable and fascinating view of antbirds and ovenbirds.

Preface

By far the richest in bird life of Earth's six major faunal regions is the Neotropical, which stretches from the northern limit of rain forests in central Mexico through Central America and over the whole of South America to Cape Horn and also includes the West Indies. Among the many peculiarly New World families found in this region, five contain over two hundred species. Three of these big families—the American flycatchers, hummingbirds, and tanagers—have expanded northward into the United States, Canada, and the Antilles; the two that are the subjects of this book, the antbirds and ovenbirds, are confined to the Neotropical mainland and a few neighboring islands. Although quite different in appearance and habits, the two are placed next to each other in our systems of classification. These passerine birds are called "suboscines" because of their simpler vocal organs and other differences from the songbirds, or "oscines." In nesting habits and the care they take of their young, they are no less advanced than the songbirds.

After more than six decades of bird study in tropical America, some of my most cherished memories are of antbirds. They have followed me for long distances through the rain forest, catching the insects stirred up by my feet or a stick. Valiant guardians of their eggs and young, they have nipped the fingers that touched their nest. They have continued to incubate while I set a camera on a tripod a yard away and photographed them. I have watched them defend their territories by striking displays that avoided fighting. In long vigils, I have admired the parents' close cooperation at their nests. Other naturalists have described how antbirds lead the flocks of mixed species that forage through Amazonian forests, warning their associates when they detected danger. Pacific birds, they live in harmony with their feathered neighbors.

With about 250 species, antbirds are the third largest family confined to the New World, exceeded in numbers only by the flycatchers and hummingbirds. More than any other great family, they are confined to wooded regions of the tropics and subtropics, rarely rising to the altitudinal temperate zone. In the ecology of the rain forests, they play important roles. Although brilliant in neither plumage nor voice, many are attractively attired in colors that blend into the deep shadows of tall forests, and their simple songs are pleasant to hear. The nests of most species are well made and faithfully attended.

The designation "antbird" is unfortunately misleading. Like many other names of New World birds, we owe it to Europeans who described lifeless stuffed skins without knowing much about the habits of the living birds. The family name, Formicariidae, is derived from *Formicarius* (from the Latin *formica,* "ant"), given in 1783 to one of its less typical but earliest described members. "Antbird" is an English approximation to this scientific name. It is appropriate only for the few species that regularly follow the army ants whose legions swarm over the ground in tropical woodlands, stirring up insects, spiders, and a host of other small creatures that are readily caught by birds while they try to escape the ants. Even the "professional" ant followers rarely eat the ants themselves. Most antbirds forage through trees and shrubs for a wide variety of small invertebrates.

Ovenbirds are an amazing family. Modestly clad in shades of brown, with only here and there a spot of spectral color, they do not attract attention by the brilliance of their attire, but their diverse tails help break the monotony of their plumage. As though to compensate for their plainness, they build a fascinating diversity of nests, including some of the most elaborate and largest made by passerine birds that do not live in avian apartment houses like those of Sociable Weavers and Palmchats. They are, above all, architects and builders. Moreover, they are extremely adaptable, living at all altitudes from seacoasts to the edges of perennial Andean snowfields, and in habitats as diverse as grasslands and tropical rain forests, marshes and arid deserts. To match the diversity of life-styles of this great Neotropical family, one must turn to a number of families of passerine birds of other lands, even to woodpeckers.

With about 214 species the fifth largest family of birds peculiar to the New World, ovenbirds are confined to South America, Central America, and tropical Mexico. None reaches the United States, where the Ovenbird belongs to the very different wood warbler family. Its domed nest of vegetable materials only superficially resembles the solid structures of hardened clay, miniatures of the domed baking ovens (*hornos*), formerly widespread in Latin America, made by several species of horneros in tropical and temperate South America. As occasionally happens in ornithology, the family has been named Furnariidae for one of its least typical genera. Only six of its species build ovens of clay; whereas scores of species make impressive closed nests of carefully interlaced sticks, often so much larger than their small builders that they might be called the birds' castles, and the family to which they belong, castlebuilders.

Whatever the name we give them, these attractive birds richly reward study of their evolution, ecology, and general behavior, including the advanced social life of some. Regrettably, they have received too little attention. In the most thorough monograph of the family, published by the American Museum of Natural History in 1980, Charles Vaurie listed 107 species—half the total number—for which he could find no or very little information on nesting. In 1983, the Asociación Ornitológica del Plata (the Argentine Ornithological Society) published *Nidificación de las aves Argentinas (Dendrocolaptidae y Furnariidae),* by S. Narosky, R. Fraga, and M. de la Peña. This small book contains detailed descriptions of the nests and eggs of the numerous ovenbirds resident in that country, information on breeding seasons, and excellent drawings of the nests, adding important information about them. Nevertheless, most of the gaps in our knowledge of ovenbirds' breeding, especially of the tropical species, remain unfilled.

Well-rounded life history studies are few; one of the most thorough that has come to my attention is that of the Rufous Hornero in Argentina by my friend Rosendo Fraga. For sixty years I have learned all that I could about nine species in Central America and one, the Rufous-fronted Thornbird, in Venezuela. The paucity of studies of ovenbirds' life histories cannot be attributed to the difficulty of finding their nests, as in the case of antbirds. In open or semiopen country, their large structures are often visible from afar, and many are not too high to be reached with or without a ladder. Unfortunately, to learn what they contain, one must often make a small hole in the wall; and no matter how carefully it is closed by the investigator and the birds themselves, such disturbance seems often to diminish the nests' success. Other ovenbirds lay their eggs and rear their young at the ends of long tunnels that they excavate in the ground, or in crevices amid rocks, where also they are not easy to see. These difficulties should challenge rather than discourage dedicated field naturalists.

Despite the large gaps in our knowledge of ovenbirds, as of other families of Neotropical birds, enough is known about their different types—the forest-dwellers and the pampas-dwellers, the castle-builders and the burrow-diggers—to undertake a wide survey of an extraordinary family of birds. By their industry, tameness, or both, some of these birds have endeared themselves to their human neighbors. I hope that this book will interest people in northern lands in two families of birds refreshingly different from those familiar to them at home, make new friends for these birds, promote their conservation in shrinking habitats, and perhaps stimulate a few adventurous spirits to undertake field studies of important avian families that have been too neglected. If it accomplishes this, I shall feel amply rewarded for the labor of writing it.

Scientific names of all birds mentioned in the text are given in the Index.

Acknowledgments

The chapter on the Black-faced Antthrush is modified from that in *Life Histories of Central American Birds,* vol. 3, published in 1969 by the Cooper Ornithological Society. The chapter on the Rufous-fronted Thornbird is taken, with alterations, from the author's article in *The Wilson Bulletin* for 1969, published by the Wilson Ornithological Society. Specimens to make the drawings were loaned by the Western Foundation of Vertebrate Zoology. The drawing of the Pale-legged Hornero's nest was based on a photograph by Manuel Marín. To all these sources of materials that helped make this book, author and artist are grateful.

Part I
The Antbirds

1 The Antbird Family

No large family of birds is more closely bound than the antbirds to the humid forests and lush thickets of tropical America. Confined to the continents and closely adjacent islands, including Trinidad and Tobago, they are absent from the United States. The species that ranges farthest north, the Barred Antshrike, barely reaches the Tropic of Cancer in southern Tamaulipas, Mexico. The whole of Mexico supports only 9 species, mostly in the wooded southeastern lowlands. As one continues southward, the number of antbirds increases. Guatemala has 11 species; Costa Rica, 30; Panama, 38.

With two exceptions, the Black-hooded Antshrike and the Streaked-crowned Antvireo, all 39 Middle American antbirds occur also in South America, where probably they originated, for this is the family's true homeland, with a vast diversity of species. Colombia supports 136; Peru, 130; Brazil, 157. Nineteen species extend beyond the Tropic of Capricorn into the woodlands and thickets of northeastern Argentina; and two of them, the Variable and Rufous-capped antshrikes, continue southward to about the thirty-fifth parallel—much farther from the tropics than they reach in North America. W. H. Hudson believed the last-mentioned, which he called the Red-capped Bushbird, to be "to some extent migratory," for he saw it only in the austral summer. Aside from this far-southern species, antbirds appear to reside permanently wherever they are found. Some, especially juveniles of certain kinds, wander locally after the breeding season.

With about 250 species the antbirds, or Formicariidae, are the third largest of the avian families confined to the New World, surpassed in numbers only by the more enterprising and widely distributed American flycatchers (Tyrannidae) with 384 species and hummingbirds

Pygmy Antwren *Myrmotherula brachyura* Sexes similar Panama to Bolivia

(Trochilidae) with 330 species. However, the family is nearly equaled in size by tanagers (Thraupidae) with 230 species and the ovenbirds (Furnariidae) with 214 species. Most abundant in warm lowlands, antbirds are still fairly numerous at middle altitudes of the Andes and other tropical mountains, while a few of them, especially hardy antpittas (*Grallaria*) reach bleak, treeless páramos around 12,000 feet (3,660 m) above sea level.

Antbirds are so called because some of them habitually forage at swarms of army ants, which they were erroneously believed to eat. Lacking popular names, ornithologists have looked to more familiar birds to facilitate the task of naming such a great diversity of species. Smallest are the antwrens, one of which, the short-tailed Pygmy Antwren, measures only 2¾ inches (7 cm) in length and is one of the most diminutive of birds. Other antwrens are not much bigger. Small-billed, delicate foliage-gleaners in lowland tropical forests where resident wood warblers are rare, they fill the ecological role of warblers in temperate-zone woodlands; antwarblers would be a more appropriate name for them. In addition to the thirty species of *Myrmotherula*, the family's largest genus, a scattering of species in other genera are called antwrens.

Antvireos tend to be slightly larger and heavier than antwrens, with somewhat thicker bills. Like the birds for which they are named, they glean insects from foliage with movements more deliberate than those of the sprightly antwrens. Antshrikes vary in size from antvireos to the biggest member of the family, the Giant Antshrike, which ranges from eastern Brazil to northern Argentina and measures fourteen inches (35.6 cm) from the tip of its bill to the tip of its tail. On antshrikes' stouter bills the terminal hook, more prominent than on other antbirds, suggested northern butcher birds, from which these antbirds differ greatly in appearance and habits. Between the foregoing groups lies a large and diverse assemblage of species known simply as antbirds, with appropriate specific adjectives. Fire-eye, bare-eye, and bushbird are names given to a few species for their idiosyncracies.

All the previously mentioned antbirds are mainly arboreal. Among the terrestrial species are the antthrushes (five species of *Formicarius* and four of *Chamaeza*), long-legged birds that walk over the forest floor with short tails uptilted, like small rails, for which they might be named. Antpittas, named for colorful pittas of the Old World tropics, are roly-poly birds with legs that seem too long for them and tails too short. The larger of them hop rather than walk over the ground, and when alarmed fly up to a low perch to look around before they jump down to bound into the nearest thicket. The smallest antpittas (six species of *Grallaricula*) fly out or down from low perches to snatch insects from the air, the foliage, or the ground, instead of hopping over

Ochre-breasted Antpitta *Grallaricula flavirostris* Sexes similar Costa Rica to Bolivia

it. The least of them are little over four inches (10.5 cm) long; but the poorly known Giant Antpitta of the Colombian and Ecuadorian Andes, and the Great Antpitta of northwestern Venezuela measure 10½ inches (26.5 cm). The bodies of these stubby-tailed birds are among the largest in the antbird family.

Also included in the family, but so different in certain features that they were until recently classified in a family of their own, are the eight species of gnateaters (*Conopophaga*), rotund, short-necked, long-legged little birds of the forest undergrowth. The males are distinguished by a long tuft of silky white feathers behind each eye, inconspicuous except when puffed out in moments of excitement.

The sexes of antthrushes and antpittas are alike, but those of many other antbirds differ conspicuously, in some cases so greatly that the male and female of the same species were originally assigned to different species. Although the plumage of these denizens of woodland shade nearly always lacks bright spectral colors, many are attractively attired in varied hues. Areas that are black, slaty, or gray on males are often rufous, chestnut, brown, or buff on their mates. Many antbirds

Black-crowned Antpitta *Pittasoma michleri* Sexes similar Costa Rica to northwestern Colombia

White-plumed Antbird *Pithys albifrons* Sexes similar Northern South America to eastern Peru

are boldly barred, streaked, spotted, or scaled. An outstandingly elegant member of the family is the male Ornate Antwren of South America. His dark gray head and neck contrast with his bright chestnut back and rump. His throat is black, his chest and abdomen gray, and his black wing coverts are boldly spotted with white. His mate is only slightly duller, with olive-brown instead of gray on her head, and white streaks on her black throat. More plainly attired males, such as the Sooty and Immaculate antbirds, are almost wholly black or blackish, with less somber females.

Ornamental plumes are rare in the family. Exceptional is the White-plumed Antbird, widespread in the forests of northern South America. Both sexes wear on their foreheads a tall, permanently erect, forked tuft of white plumes, balanced by a white "beard" on the chin. The head is otherwise black; the back and wings dark blue-gray; the nuchal collar, rump, tail, and underparts chestnut. The eyes are yellow. Bushier crests adorn the heads of the Black-crested Antshrike, his

Reddish-winged Bare-eye *Phlegopsis erythroptera* Male Venezuela to eastern Peru

chestnut-crested mate, the Tufted Antshrike, and the Chestnut-crested Antbird. Grayish white filaments cover the crowns of both sexes of the Hairy-crested Antbird, who often hold them flat.

When a male Barred Antshrike raises the long black feathers of his crown, their white bases attract the eye. In other antbirds, dark feathers in other places have light bases that are hidden except when these feathers are raised or spread outward in territorial disputes, nest defense, and other occasions of excitement. Then they form gleaming white patches in the center of the back of the Great Antshrike, Slaty Antshrike, Dusky Antbird, Spotted Antbird, Dotted-winged Antwren, and others; on the shoulders of the Plain Antvireo and Slaty Antwren; or on the forward edges of the wings of the Chestnut-backed Antbird. The Russet Antshrike's hidden back patch is cinnamon-rufous. In a number of species, including the Slaty Antshrike and the Dotted-winged Antwren, both sexes have concealed areas of contrasting color. All these antbirds expose their white or buffy feather bases in such circumstances as incite American flycatchers to spread the feathers of their crowns and display spots of red, yellow, or white that most of the time are well covered.

The eyes and bare skins of antbirds often have colors that their plumage lacks. The irises of the Great Antshrike and the fire-eyes

(*Pyriglena*) are intensely red, those of the Fasciated Antshrike and the Black-capped Antshrike of a redness less brilliant. Bright yellow eyes stare out of the heads of Barred Antshrikes and some related species, but dark eyes are more widespread in the family. The orbits of a number of antbirds are surrounded by bare skin, which may be bright red, as in the bare-eyes (*Phlegopsis*); bluish white, as in the Hairy-crested and Chestnut-crested antbirds; or brighter blue, as in the Bicolored, Ocellated, Immaculate, and Chestnut-backed antbirds. While closely watching the last-mentioned at a nest, I was impressed by the mobility of their eyes and the way the skin around them contracted on the side toward which they looked, expanded on the opposite side. Bare skin should be more flexible, and less obstructive to the oblique glance, than feathered skin. To have naked skin around their eyes may be advantageous to birds that hunt their food amid vegetation that presses close around them. Moreover, bare surrounding skin that accentuates the eyes, or at a distance makes them appear bigger, may deter certain predators, as do the large eyespots of certain moths, butterflies, and other insects.

The bright blue of the naked skin that covers the whole face, forecrown, and forehead of the black male Bare-crowned Antbird appears

Black Bushbird *Neoctantes niger* Male Colombia to Peru and Amazonian Brazil

Barred Antshrike *Thamnophilus doliatus* Male Mexico to northern Argentina

to be the greatest expanse of this color in the whole antbird family. On the brown female, only the orbital skin and lores are bare and blue.

A feature unusual in the family is the chisel-like bill with upcurved lower mandible of the Black Bushbird, who uses it to dig into decaying fallen logs, in the manner of a small woodpecker. Stranger still is the laterally compressed bill of the Recurved-billed Bushbird, with a more strongly upswept lower mandible, employed for splitting stems of heliconias and other herbaceous monocotyledons.

Antbirds are most numerous in humid forests and thickets. In tall rain forests, most of them avoid the sun-bathed canopy, through which more brilliantly attired birds flit, and remain in the shade from about midheight of the great trees through the leafy lower strata, where antwrens and antvireos mostly forage, down to the ground, over which antthrushes walk and antpittas hop, and to which arboreal species occasionally descend. The riotous growth of saplings, shrubs, and creepers in sunny openings made by fallen trees offers the seclusion and abundance of small invertebrates that many antbirds prefer. The lush thickets that soon cover abandoned clearings and resting fields in rainy regions offer food and shelter to a few antbirds that forage low. Although woodland cotingas, flycatchers, woodcreepers, woodpeckers, and other birds frequently emerge from the forest to forage and nest in nearby clearings with scattered trees and shrubs, antbirds consistently avoid such sunny places. The few antbirds of arid country inhabit gallery forests along rivers and the moister streamside thickets.

As far as is known, all antbirds are monogamous and live throughout the year in pairs or family groups, mostly on defended territories. Although antthrushes and antpittas forage alone, they apparently remain mated. Like other sedentary birds, antbirds seldom gather into flocks of a single species, but they are often conspicuous members, sometimes the leaders, of the mixed-species flocks that forage in tropical woodlands. Typically, such an aggregation contains only a single pair or a single family of any one species of antbirds, who noisily refuse admission to others of their kind but offer no resistance to compatible birds of other kinds, as will presently be told.

One of the most adaptable and familiar of antbirds is the Barred Antshrike, which ranges from Mexico to northern Argentina and inhabits arid as well as rainier country; it is the most abundant antbird on the semiarid Pacific coast of Middle America. Less reluctant to expose itself in the open than most antbirds are, on the island of Tobago it visits houses for soaked bread, a surprising food for insectivorous antbirds, and in Venezuela it also frequents feeding stations. A pair sometimes foraged in the garden of a large farmhouse that we occupied in Venezuela, usually coming early on warm afternoons, when the house was quiet and nobody was in sight, and revealing their presence by a surprising variety of utterances.

In Central America, I have found Barred Antshrikes more retiring. How reluctant they are to abandon the thickets where they lurk was vividly demonstrated to me in a Costa Rican valley one afternoon at the end of a dry March, when such a thicket was burned to clear land for planting without the usual preliminary slashing down. As with loud crackling and dense clouds of smoke the flames spread through the acre or so of thicket, mobile flycatchers and tanagers fled well in advance of the blaze. Pigeons and doves followed soon after. Finally, even a secretive Plain Wren could endure it no longer and rushed forth to fly slowly and laboriously over an adjoining open field. But three Yellow-billed Caciques and a pair of Barred Antshrikes clung tenaciously to the burning thicket. Reluctantly retreating close ahead of the conflagration, they were finally driven into a corner, where they defied the heat and smoke. Fortunately for them, this small patch of

Rain forest in Almirante Bay region of western Panama, habitat of Slaty Antshrike, White-flanked Antwren, and Chestnut-backed Antbird

bushes, only a few yards wide, remained unburned. Here the two antshrikes stayed with the three black caciques.

Once I watched a Bicolored Antbird cross a pasture by flying low between scattered trees and bushes. This was a surprising excursion, for as I shall presently relate, I could not entice into the open a Bicolored Antbird who followed me for long distances through the forest, catching the insects I stirred up for him. Antbirds' aversion to exposing themselves or to crossing open spaces appears to be in part responsible for the great diversity of species in South America. The wide rivers that dissect the vast lowland forests of Amazonia and the Guianas, the high ridges that separate deep Andean valleys are barriers that, by isolating sedentary populations, favor the evolution of new races and species. A respected hypothesis holds that during drier intervals of the Pleistocene glaciations Amazonian forests were restricted to moister areas called "refugia," separated by expanses of savanna or arid scrub that antbirds would avoid. Such fragmentation of the humid forests, if it actually occurred and continued long enough, would also increase the number of species of woodland birds. However this may be, whether transported by winds or their own efforts, antbirds have over the ages transcended barriers, until today many species are very widespread, from the Caribbean countries to the Tropic of Cancer, while others have continued to abide in narrow regions. But unlike certain other birds of South American families that are less averse to open country, notably flycatchers and hummingbirds, no antbird has crossed the wide Caribbean to become established in the Antillean islands. Trinidad, with ten species of antbirds, is not Antillean but a slightly separated piece of South America.

2 *Food & Foraging*

Antbirds are insectivorous and rarely take fruits. Antwrens flit, warbler-like, through low and middle levels of the forest, gleaning small insects and spiders from foliage and twigs. By painstaking observation, Russell Greenberg and Judy Gradwohl learned that White-flanked Antwrens, Dotted-winged Antwrens, and other small birds such as the Spotted-crowned Antvireo and the Slaty Antshrike gather many more insects from the lower than from the upper surfaces of leaves. Small flycatchers of tropical woodland, as well as manakins and other birds, pluck insects chiefly from beneath the foliage, where inside the forest they are more abundant than on the upper faces. Moreover, a bird perching on a leafy twig has a less obstructed view of the twigs above itself than of those below, which are often screened by the foliage at the level of its feet, and it can more readily dart upward to pluck items from beneath leaves while it flies past or hovers below them.

Although Checker-throated Antwrens accompany White-flanked and Dotted-winged antwrens in mixed-species flocks and sometimes forage with them amid the boughs of smaller trees, up to thirty feet (9 m) above the ground, mostly they remain lower, scrutinizing the herbage and investigating the foliage of ferns down to their roots. They find many small creatures in curled dead leaves that have lodged in the vegetation near the ground. Often they bite along the folds of withered leaves to force out whatever caterpillar or spider might be lurking within. I have not seen Checker-throated Antwrens actually hunt over the ground, although they are much easier to keep in view than are the antwrens that forage at higher levels, moving rapidly through screening foliage.

Dusky Antbirds forage through densely entangled vegetation, like many tropical wrens which they resemble in size and form of bill and body. They climb among twisted vines and clustered living or dead leaves, searching assiduously for spiders, caterpillars, moths, and insects of many other kinds. The larger antshrikes flit and hop more heavily through thickets, capturing an occasional small lizard or frog in addition to the arthropods that they mainly eat. Sometimes a few berries add variety to their fare, as has been recorded of the Black-crested, Black-hooded, and Barred antshrikes. As already noticed, on Tobago the last-mentioned comes to feeding tables for soaked bread, a habit which may have begun after the hurricane of 1963 made other foods scarce on the island. The Giant Antshrike and the big Large-tailed Antshrike eat eggs and nestling birds, small rodents, and snakes.

Birds that do not habitually forage on the ground may descend to it to retrieve an item that they have dropped, or to capture a creature that has attracted their notice. One afternoon I watched a female Black-hooded Antshrike hop down to the ground beneath a thicket, where she made fallen leaves fly, vigorously tossing them right, left, and upward with her strong bill. When she uncovered and seized a large green orthopteron that had sought refuge in the loose litter, violent movements of the insect's wings and legs made her jump up as though stung. Again and again she renewed the attack, only to withdraw when her intended victim defended itself with vigor. But the bird stubbornly persevered, until by intermittent pecks or bites she finally subdued it, seized it in her bill, and flew away with her prize.

A female Dusky Antbird was more easily deterred from an intended meal. At the forest's edge, she found a big, hairy brown spider clinging to a leaf, and from a cautious distance she stretched her neck forward to seize it in the tip of her bill. The spider raised its front legs with the two on each side in contact except at their tips, which were slightly separated and resembled a scorpion's pincers. Whenever the bird came near, the spider waved these uplifted legs, making her draw back. Soon she desisted from her attempt to seize it and departed, leaving it to creep into its retreat between two leaves that it had fastened together. Measuring about two inches (5 cm) across its spread legs, it was a big creature for a small bird to tackle.

A semiterrestrial forager is the Chestnut-backed Antbird of Central America and northwestern South America. Over more open ground beneath humid forest it hops rapidly instead of walking, picking up small edible items. Occasionally with its bill it tosses aside a leaf beneath which it has seen a small creature seek refuge. Where the ground cover is dense, it advances by flitting from stem to stem. Much of its food is found amid vine tangles, on trunks wreathed in creepers or covered with moss and larger epiphytes, among dead leaves lodged in the vines, and between old fronds that drape the stems of small palms. Sometimes a Chestnut-backed Antbird works upward through tangled vegetation to a height of ten feet (3 m), but usually these birds stay lower while they search for invertebrates, small lizards, and tiny frogs. Occasionally, they forage with army ants.

Antpittas appear to find most if not all of their food on the ground, over which they hop, at intervals tossing aside fallen leaves with their bills. Nearly always an antpitta is alone. A Spectacled Antpitta is amusing to watch as it stands motionless on the dimly lighted, leaf-strewn forest floor, its stout, nearly tailless body propped up on legs so long and slender that they look like thin sticks, the light rim around each dark eye giving it a startled expression. As though it were not already sufficiently plump, it rhythmically puffs out and compresses the streaked feathers of its white breast. At intervals it half spreads and closes its short brown wings—a movement which, like the wing-spreading of mockingbirds, may stir up insects. Some Brazilian antpittas are reported by Helmut Sick to eat seeds, and to be captured in traps baited with grain.

Antthrushes appear more thoroughly adapted for terrestrial foraging than antpittas are, for they walk instead of hop over the ground, pushing or throwing litter aside with their bills. To the insects and other arthropods that they uncover they add berries, seeds, snails, small lizards, frogs, and snakes. The Short-tailed Antthrush of Brazil is one of the many birds of diverse orders that eat the dark, juicy berries of pokeweed (*Phytolacca*) that springs up abundantly on burnt or otherwise denuded land. Once, while I sat in a blind in the forest, watching a Ruddy Quail-Dove's nest, my attention was drawn by the clear whistles, followed by sharp notes, of a Black-faced Antthrush. Presently, through a side window, I glimpsed the antthrush struggling with a snake nearly a foot (30 cm) long, which it tried to subdue by pecking and knocking while it writhed on the ground. After this had continued for some minutes, a second antthrush walked hurriedly up, possibly to receive the serpent from its mate. The outcome of this encounter was hidden from me by the foliage into which the birds promptly vanished.

Although, like many other birds, antbirds occasionally catch in the air insects that pass temptingly near, only a few are habitual flycatchers. Among them are the Cinereous Antshrike and other species of *Thamnomanes,* who sit upright on slender horizontal branches within a few yards of the ground, from which they sally to snatch insects from the air or foliage, returning to the same lookout, in the manner of many American flycatchers. From lower stands, Rusty-breasted Antpittas dart out to catch small creatures in the air, from foliage, or on the ground, as told in the preceding chapter. Gnateaters forage in much the same way. Exceptional among antbirds is the foraging of the Black Bushbird, which pecks into decaying wood for food, and that of the Recurved-billed Bushbird, which splits open soft stems.

Foraging with Army Ants

The forests of tropical America are famous for the multiplicity of their birds, but the population of any single species tends to be sparser than that of birds of meadows and open marshes. Probably because they are stimulated by fewer near neighbors and rivals, birds of tropical forests are less songful than those of northern lands. One may wander long through tropical woodland, disappointed by the fewness of birds visible or audible, perhaps oppressed by the silence and paucity of evident animate life amid profuse vegetation, until he or she meets a concentration of birds that appears to include most of the feathered inhabitants of a wide surrounding area. These birds may have gathered at a generously fruiting tree; they may be a flock of diverse species foraging through the treetops or lower levels, or they may accompany a swarm of army ants. Only the last two of these assemblages are likely to include antbirds.

One's first intimation that ants are swarming is often a chorus of protesting churrs, rattles, and trills from birds amid the undergrowth whose feasting has been interrupted by the inadvertent intrusion. His attention held by a multitude of flitting forms, the intruder may not notice that he has blundered into the midst of a legion of hunting ants until a few stings on his legs make him look down. The ground around him is covered with hurrying ants, darkening the leaf litter as they swarm over and beneath it, while columns of them file up surrounding trunks, often until lost to view amid the foliage above. If rain has not recently fallen, the movements of myriad bustling ants and small creatures trying to escape them audibly rustle the dry leaves.

As the ants approach, their potential victims scurry from beneath concealing ground litter and crevices in bark; concealing coloration or cryptic habits avail them little now. Roaches, crickets, spiders, millipedes, sowbugs, scorpions, frogs, and lizards—everything able to move—expose themselves recklessly in a frantic attempt to save their lives that is nearly always futile, for those that escape the ants fall easy prey to the attendant birds. This is their opportunity; foraging with army ants, they avoid tedious rummaging through leaf litter, scanning foliage, or searching over mossy, vertical trunks. The ants stir up prey for them much as grazing quadrupeds do for cowbirds, anis, and Cattle Egrets in open pastures. By foraging with ants, they can fill their stomachs much more rapidly than when hunting without them. Small birds that follow army ants do not, except perhaps incidentally, add these ants to the great variety of other creatures that the ants make available to them; larger birds, such as domestic chickens, may devour the ants at the forest's edge or beyond it in more open places.

Through the length of tropical America and beyond it, from northeastern Mexico to northern Argentina and upward to about 6,000 feet (1,830 m) in the mountains, the ant that attracts the largest gatherings of birds is *Eciton burchelli,* a middle-sized species with soldiers half an inch (13 mm) long and workers down to a third this length. They vary in color with race and locality but are mostly blackish. Dorylines related to the dreaded driver ants of Africa, their hordes are smaller and less formidable. They have strong grasping mandibles, and stings about as painful as those of much smaller fire ants. Lacking permanent abodes, they pass the night in bivouacs that are among the most fantastic sights of tropical forests. In a hollow log or trunk, a hole in the ground, beneath a stout liana or an overhanging rock, amid close-set branches of a shrub, in a woodpile, or some similar location, at ground level or high above it, hundreds of thousands, sometimes a million, ants cling together by hooks on their legs in a compact ball up to a foot in diameter, the size of a basketball or larger. In the center of this tissue composed not of cells but of separate individuals capable of independent movement, the single queen and her young brood find protection in an equable ambience.

Throughout the year, in cycles lasting about five weeks, the ants alternate between a statary and a nomadic phase. This cycle is related to the stages of development of two overlapping broods and the queen ant that produces them. As the larvae of one brood, which the workers have been carrying from nightly bivouac to nightly bivouac, begin to spin their cocoons, nomadism ceases and the statary phase begins. Now the queen, who has been slender and agile enough to walk with the horde from nightly encampment to nightly encampment, is so abundantly fed by workers without larvae to nourish that her abdomen swells until the chitinous scutes that covered it separate, exposing the underlying skin. In this "physogastric" state, she becomes a machine that can produce a hundred thousand or more eggs in a few days, after which her abdomen shrinks.

About the time these eggs hatch, the ants of the preceding brood emerge from their cocoons; the empty cases cover the ground beneath the statary bivouac, looking like small whitish beans. Now, with up to a hundred thousand larvae to feed, and a work force augmented by as many thousands of callow ants to nourish them, the colony abandons the bivouac that it has occupied continuously during a statary phase of about three weeks' duration and embarks upon a nomadic phase that will last eleven to sixteen days. During this interval of wandering, the encampment is on most nights shifted for distances up to 850 feet (260 m). This cyclic alternation of statary and nomadic phases has important consequences for attendant birds. During the former, when the ants need less food, daily raids tend to be weak, of short duration, or sometimes omitted; during the latter, when many thousands of rapidly growing larvae must be fed, raids become more vigorous and dependable and attract more birds.

A strong raid usually starts at dawn, when the ball of tightly crowded ants dissolves and they spread over the surrounding ground. Soon the dispersed ants consolidate in a lengthening column several individuals wide. As it advances, the column sends out spurs to the right and left. Presently, it splits into branches that anastomose like the channels of a great river at its delta. Finally, the branches unite in a single seething tide of ants that on a front up to fifty feet (15 m) wide flows forward over the forest floor. In this advancing front many thousands of ants scamper about in all directions, seeking insects and other small invertebrates. With the insect's compound eye reduced to a single ocellus, they are nearly or quite blind but sensitive to light;

they must find their victims by scent or simply by bumping into them. Each larger captive becomes a seething mass of dusky ants struggling to overpower and tear it apart. Back along their trail they carry their booty past other ants hurrying outward to join the fray. Usually their foraging ends and the troops march toward their bivouac in the late afternoon or evening, but if they have captured a large wasps' nest, they may continue far into the night to carry out the larvae and pupae that they tear from the combs.

These ants are not nearly as formidable as sensational accounts depict them. They specialize on invertebrate prey. Active vertebrates of all kinds readily avoid them, perhaps not without a few stings. The ants' mandibles are not sharp enough to dismember them even if they could overwhelm them, as might happen if the animals were already moribund or trapped, as in a pit. Often I have continued to sit at my table and write, while army ants scurried over the floor around me and the ceiling above me. I have watched them flow past nests on which birds ranging in size from tiny seedeaters to Great Tinamous sat, without harming them. When scouts from a swarm reached a low nest where a Spectacled Antpitta brooded newly hatched nestlings, he calmly picked them off the rim and ate them.

Army ants like *Eciton hamatum,* which raid in columns instead of deployed battalions, do not attract birds. After *E. burchelli,* their chief purveyor of food is *Labidus praedator,* a noticeably smaller reddish brown or black ant. From its subterranean abode it emerges, chiefly when the forest floor is wet or damp after recent rain, to spread over it in dense sheets rarely more than ten feet (3 m) wide, from which detachments may climb up through the vegetation to about fifteen feet (5 m). These smaller ants seek smaller, less active prey, including sowbugs and amphipods, and after a few hours of activity they flow back into the ground whence they came. They rarely attract such large crowds of birds as forage with *Eciton burchelli.*

On Barro Colorado Island in Gatún Lake, through which the Panama Canal passes, army ants and their followers have been intensively studied, first by Robert Johnson and later, for many years, by Edwin Willis. Here colonies of *E. burchelli* occur at a density of about eight per square mile (3 to 3.2 per square kilometer), or one for every eighty acres (33 hectares). This is a large tract of forest through which to seek them. Antbirds find an active horde by waiting by the previous night's bivouac until the ants emerge, by hearing or seeing other birds already foraging with the ants, or by searching through the area where the ants were last found. A person may locate an active swarm by similar methods, or simply by chance while wandering through the forest. By whatever means one finds a swarm, he beholds an unforgettable scene of concentrated activity. Birds that withdrew when he first appeared resume their foraging if he stands quietly to watch at a convenient distance. The species of attendant birds differ from region to region over the vast territory where army ants hunt and birds accompany them, but everywhere the spectacle is much the same.

In southern Central America, habitual or "professional" ant followers include, first and foremost, Bicolored Antbirds, of which I have counted at a swarm from one or two up to ten or twelve, including immature birds, with probably more lurking unseen in the background. Next in prominence come little Spotted Antbirds and family groups of large Ocellated Antbirds. Occasionally a Bare-crowned Antbird joins the throng. At altitudes higher than heat-loving Spotted Antbirds reach, a black Immaculate Antbird and his deep brown mate may be present. Surrounding these birds that occupy center stage, many individuals of other families are nearly always active. One or more Gray-headed Tanagers, shier than the antbirds, hover on the outskirts, nervously twitching their wings and tails. Plain-brown, Tawny-winged, or Ruddy woodcreepers, more rarely a big Barred Woodcreeper, cling to trunks above the swarm; tiny manakins, mostly females, and small flycatchers flit through surrounding shrubs and saplings; occasionally a Great Tinamou or a Black-faced Antthrush walks slowly on the outskirts, picking up fugitives. Migratory Swainson's Thrushes forage with them.

Even a raptor, such as a Barred Forest-Falcon, may be attracted to the scene, apparently for food other than birds, which it is not known to catch at these gatherings, although elsewhere it preys heavily on them. For two hours I watched an immature Collared Forest-Falcon foraging with a motley crowd of smaller birds, who stayed at a respectful distance from this larger raptor while it caught big insects and spiders without, as far as I saw, menacing any of the other birds.

Each attendant at ant swarms operates in its own way. Woodcreepers pluck creeping things from trunks, or fly out or down to snatch them from the air or the ground, then return to their vertical stations; manakins pick insects from leaves or stems as they dart by; flycatchers seize them in the air. Antbirds cling to slender upright stems of

Bare-crowned Antbird *Gymnocichla nudiceps* Male (above) and female (below) Belize to northern Colombia

saplings, rarely more than a yard up, wagging their tails up and down while they scrutinize the ground, to which they drop, capture a small fleeing creature they have spied, then immediately rise with it to a sapling. Bicolored Antbirds subdue a victim, and dismember a larger one, by chewing and shaking it rather than beating it against their perch. Spotted Antbirds more often pound their prey.

The rapid ascent of an antbird that has captured an insect amid ants swarming over the ground suggests that it fears the hunters that it has deprived of a victim, usually before they have attached themselves to it. Although the bird tries to avoid becoming overrun by the ants, it is in little danger from them. I have watched army ants filing along a stem or vine on which a Bicolored Antbird rested. The insects either turned back, passed around the bird's toes, or sometimes over them, without making the bird move. Ocellated Antbirds perched calm and unperturbed while army ants flowed over their feet. Likewise, a Barred Woodcreeper, clinging to a trunk along which ants were passing, crawling over its toes and even its plumage, did no more than from time to time pluck one of them off with its strong bill. When a Pale-faced Antbird stands on the ground amid army ants, as it seldom does, it jiggles from one foot to the other, or frequently leaps into the air. In Nicaragua a collector shot a Fasciated Antshrike that fell in front of advancing army ants which soon surrounded it, preventing him from retrieving the bird. When he returned the next day, expecting to gather up a skeleton from which the ants had removed the flesh, he found the corpse intact. The ants had not even attacked the bare skin around the bird's eyes. A freshly killed snake thrown into the midst of hunting ants also was ignored by them.

Days may pass before one of the widely scattered legions of army ants enters the territory of a given pair of Bicolored Antbirds. While waiting for this to occur, these birds so dependent upon the ants to purvey food to them might starve, and their young would perish. To avoid this disaster, Bicolored Antbirds have evolved a loose and flexible territorial system, which Willis elucidated. Each pair is dominant in its own territory, but when army ants happen to enter, it permits other Bicolored Antbirds to forage with them, always in a subordinate role. Pair A asserts its priority over pairs B, C, etc., by frequently supplanting them on their perches, without fighting or trying to expel them from territory A as long as the ants continue to hunt there. When the legion moves to territory B, or C, pair A may follow it there, now becoming subordinate to whatever pair claims the area. This system led Willis to define territory as "a space in which one animal or group dominates others which become dominant elsewhere." The Bicolored Antbird's tolerance of intruders of its own kind appears to be indispensable for the survival of the species while its present foraging habits persist, but it is incompatible with the widely held view that the individual does nothing "for the good of its species."

With stronger territoriality combined with the smaller size that makes them subordinate to larger ant followers, Spotted Antbirds forage less consistently at ant swarms but are better able to procure food

Ocellated Antbird *Phaenostictus mcleannani* Sexes similar Honduras to northwestern Ecuador

away from them, as they do about half the time, alone, in pairs, or with mixed-species flocks of birds.

In the wet forests of the Andean foothills of eastern Ecuador, I watched pairs of White-plumed, Spotted-backed, and Scaly-backed antbirds, with a Spotted Nightingale-Thrush, foraging with army ants. Unlike other antbirds, long-toed Scaly-backs can cling horizontally to thick trunks. Among the twenty-eight antbirds known to be "professional" ant followers are also five species of *Rhegmatorhina,* distinguished by wide rings of bare bluish skin around their eyes: the Bare-eyed, Harlequin, Chestnut-crested, Hairy-crested, and White-breasted antbirds. Two species of *Phlegopsis,* whose eyes are set amid broad areas of red skin—the Black-spotted and Reddish-winged bare-eyes—also regularly accompany ants. Three relatives of the Bicolored Antbird in the genus *Gymnopithys*—White-throated, Rufous-throated, and Lunulated antbirds—are likewise "professional" ant followers.

Foraging with an Ornithologist

Rarely, an antbird uses a human instead of army ants to stir up insects. On a morning in February, long ago, I paused in my walk along a woodland trail, on our farm in southern Costa Rica, to search for a

Bicolored Antbird *Gymnopithys leucaspis* Sexes similar Honduras to northeastern Peru and northwestern Brazil

nest in a hollow palm stump beside which was a Bicolored Antbird. While I examined the stump, the bird watched me closely. When, finding no nest, I resumed my walk, the antbird followed me, snatching up insects driven from fallen leaves by my feet. Moving slowly, taking care to stir the ground litter and to avoid abrupt movements, I led it through the undergrowth for about a hundred yards, a leisurely journey on which it was rewarded with many small creatures that I made available to it.

This was the first of many similar excursions that the antbird and I took together over the next sixteen months. Although—the sexes of Bicolored Antbirds being alike in appearance—I was not sure that a masculine name was appropriate, I contracted *Gymnopithys* to "Jimmy," and when we were together I repeated this to my woodland friend to accustom him to my voice. Sometimes I found him in the forest; more often he found me, sometimes while I stood watching some other bird. Apprised of his presence by a low, questioning note, I would look down and see him clinging to a slender, upright sapling, a foot or two above the ground and hardly a yard away, his lower leg stretched and his upper leg flexed to hold his body almost horizontal. If I delayed to move, he quietly preened, at intervals repeating a little throaty note to remind me that he was waiting to be served. Even after an interval of two or three months during which we had not met, my little friend would come to me as though we had parted only yesterday. As soon as I started on a walk, he followed.

Instead of with my foot, I often stirred the ground litter with a short stick. When a suitable insect was exposed, Jimmy would dart out and seize it with a *clack* of his black bill. When he saw an escaping creature dive beneath concealing leaves, he would alight on a low twig or the ground and flick them aside with his bill, just as antthrushes, antpittas, and many other ground-feeding birds do. He had definite preferences and made no move to capture certain insects that to me looked no less appetizing than those he eagerly devoured. Among the creatures he disdained were certain moths; but he ate other kinds of moths, gulping them down with wings still attached. Although he did not knock invertebrate prey against his perch, as many flycatchers do, he used this procedure to immobilize small frogs that I made available to him, mostly in the rainy season. While pursuing a fugitive insect he would sometimes fall several yards behind me, but soon, flying low above the ground, he would return to my side. When I delayed too long in one spot, he would prompt me to proceed by repeating his low, throaty note while he clung to a sapling close beside me.

Although Jimmy caught insects beside my feet, he would never permit my hand to approach closer than a few inches from him. However, with the short stick that I used to stir the litter, I could ruffle his plumage or touch his bill. He continued to cling to stems that I rudely shook, and the loudest noises that I could make did not perturb him at all. He seemed fearless of everything except the human hand—that strange, grasping organ carefully avoided by many animals that otherwise permit a close approach by man.

The part of the forest where mostly I met Jimmy was about a thousand feet (300 m) across. One day I led him far beyond this, for about half a mile, which seems a long distance through undergrowth. When I returned by a little-used wood road, he refused to accompany me. Apparently, he had no difficulty finding his way home alone, for a few weeks later I found him on familiar ground. When I tried to entice him from the forest into an adjoining sunny pasture, where I might have stirred up fat grasshoppers for him, he stubbornly remained amid the dense vegetation at the woodland's edge. He would accompany me into shadier parts of tall second-growth woods beside the old forest, but he avoided more open spaces where spots of sunshine were more concentrated on the ground. Like other antbirds, he was definitely photophobic. Nevertheless, Bicolored Antbirds have followed army ants into our garden adjacent to the forest, keeping in the shelter of hedges and shrubbery.

Although Bicolored Antbirds live in pairs throughout the year, Jimmy was nearly always alone. Possibly he had left his mate to come and feast upon the insects I procured for him. On rare occasions I glimpsed, lurking in the background, a Bicolored Antbird that I surmised to be his consort, but usually it faded away and we two would go on alone. When a second antbird approached us too closely, with a growl Jimmy would warn it to keep its distance, or fly toward it as though to chase it away. He evidently regarded me as his private servitor, and did not show the same tolerance of commensals of his species as Bicolored Antbirds commonly do at ant swarms.

One day, the second antbird who had followed obscurely at a distance, gaining confidence, approached us more closely until it foraged profitably only eight or ten yards away. Apparently, it was near enough to find insects that had been disturbed by my passage and had not yet concealed themselves well. Jimmy did not again fly at the other bird, or repeat his warning growls. Although in the past he had always followed me in silence, save for these notes and the slight, confidential utterances that he used to attract my attention when I neglected my self-imposed role of purveyor, now he and the other bird communicated with their little laughing song of low, short notes running up the scale. This strengthened my belief that the two were mates, but it made me doubt that I had attributed the correct sex to Jimmy. Male antbirds often feed their mates; if indeed of this sex, Jimmy should have been more generous. Moreover, Willis noticed that male Bicolored Antbirds are frequently henpecked by dominant partners. Male or female, Jimmy was the free, undomesticated animal with whom, over the years, I have been most intimately associated. His or her confidence in me was deeply gratifying and has left a cherished memory.

After a year and a half, during which we had spent uncounted hours together, Jimmy passed from my ken. At intervals over nearly half a century, other Bicolored Antbirds have foraged with me much as Jimmy did, and might have become as intimate if I had devoted more time to them. More surprisingly, one morning when with a man and a boy we cut a trail through the forest, a Bicolored Antbird closely followed the three of us, and caught insects stirred up by the boy's swinging machete.

On another occasion, I met along a forest path a pair of antbirds who seemed so tame that I encouraged them to follow me. After we had advanced several hundred yards together, while they ate the food I made available to them, we came in sight of two other Bicolored Antbirds. They and my two companions started to chase one another; apparently, we had invaded the territory of the second pair, who did not hesitate to assert their ownership. After this, the first pair would follow me no longer; but two months later I was again accompanied on a walk through the forest by two antbirds, doubtless the same couple. Like Jimmy, all these companions would permit me to touch them with a stick, and to come within a few inches of making contact with a hand. Possibly they were Jimmy's descendants. Although occasionally an ant-following bird accompanies quadrupeds, such as a herd of peccaries, no other antbirds of any species have anywhere foraged with a human, as far as I know. Even Willis, who spent about three thousand hours studying antbirds at ant swarms on Barro Colorado Island, never had this experience.

Young Bicolored Antbirds, with dark patches on their white breasts and yellow mouth corners, have from time to time watched me intently while I stood in the forest, from perches less than a yard away. When I moved, some of these young birds followed, then paused to gaze at me from another upright stem. Sometimes, like the adults, a juvenile permitted me to advance a hand to within ten or twelve inches, but never to touch it. I could never induce one of these immatures to forage with me; the parents, who were nearby and probably still fed them, were a stronger attraction.

3 Daily Life

When not courting, attending nests, or feeding dependent young, antbirds spend their days alone, in solitary pairs, or family groups, or as individuals, pairs, or family groups in mixed-species flocks. Antthrushes and antpittas are usually alone, antshrikes (*Thamnophilus*) and antvireos (*Dysithamnus*) alone or in pairs, and many arboreal antbirds in flocks of diverse species. The strong attachment of antbirds to territories and their sedentary habits are not compatible with the formation of large flocks of a single species. The greatest aggregation of antbirds of the same species that I can recall having seen in many years amid the forests where they dwell was at most a dozen Bicolored Antbirds at a swarm of army ants. They were not a true flock or social group but a gathering of families or individuals permitted by Bicolored Antbirds' relaxation of territorial exclusiveness and tolerance of intruders at ant swarms, as explained in the preceding chapter.

Many antbirds forage in flocks of antbirds of several species and birds of different families. In the forest on Barro Colorado Island, I found flocks of antwrens, which in later years were more carefully studied by R. Haven Wiley and by Russell Greenberg and Judy Gradwohl. These flocks were composed mainly of pairs, often with their young, of Checker-throated Antwrens, Dotted-winged Antwrens, and White-flanked Antwrens. The first two species shared territories that averaged 3.7 acres (1.5 hectares). In the dry season, when they foraged more widely than in the wet season when food was more abundant, they ranged over about 8 acres (3.2 hectares). Territories of the White-flanked Antwrens were larger and more variable in size, ranging from 3.7 to 11 acres (1.5 to 4.5 hectares) and averaging 6 acres (2.5 hectares). They often overlapped two or three territories of the other two antwrens. The territories of Checker-throats and Dotted-wings were stable; their boundaries remained essentially unchanged during the seven years that Greenberg and Gradwohl studied these birds.

As a rule, antbirds and others stay with a wandering mixed-species flock only as long as it remains in their own territories. When the flock crosses a territorial boundary, the residents of that territory drop out but are often replaced by birds of the same species in the adjoining territory—a substitution that may be overlooked if the birds are not individually recognizable, as by colored leg bands. For this reason, White-flanked Antwrens, with larger territories, are not constantly associated with the same Dotted-wings and Checker-throats, as these two are with each other. When two or more species claim the same area, each defends it only against others of its own kind: Checker-throated Antwrens against Checker-throats; Dotted-wings against Dotted-wings; males against males; females against females. Antbirds do not proclaim possession of territories by profuse singing, as many northern birds do, but defend their domains by displaying and chasing at the boundaries, seldom fighting.

The boundary disputes of Dotted-winged Antwrens are frequent and often prolonged, continuing from a few minutes up to three hours. Not only do adults of both sexes participate in these confrontations, posturing and exposing their white back patches, but juveniles and immatures may join them, or sing in nearby trees. In contrast to Dotted-wings, female Checker-throats often continue to forage while their mates display, chirping antiphonally with their counterparts across the border. Likewise, female and immature White-flanked Antwrens rarely participate in border disputes, which are largely exchanges of calls by rather widely separated males, followed rarely by swift, long-distance chases.

The three species of antwrens in these flocks forage through lower levels of the forest, up to 30 or 40 feet (9 to 12 m) above the ground, in different ways. Both Dotted-wings and Checker-throats prefer vine tangles, but while Dotted-wings glean insects from green foliage, Checker-throats rummage among dead leaves, as told in chapter 2. White-flanks flit, warblerlike, from twig to twig through more open spaces, gathering insects and spiders. Thus, the three closely associated antwrens do not compete seriously for food, but they complement one another in several ways. By exposing their white flank plumes when they spread their wings in flight, black White-flanked Antwrens provide visual signals that may help to guide the mixed flock when it deploys through about 100 feet (30 m) of screening foliage. With louder calls, more freely given, while they mob predators or intruding bird-watchers, Checker-throats warn the flock of potential danger. Other birds that accompany these antwrens for longer or shorter intervals include Slaty Antshrikes, Plain Xenops, and Plain-brown Woodcreepers.

Amid the very rich avifauna of Amazonian forests at Cocha Cashu Biological Station in Manu National Park in southeastern Peru, Charles Munn and John Terborgh followed larger flocks of antbirds and associated species of other families. The core members of the flock they watched most continuously were twelve species which shared the same twenty-acre (8 hectare) territory. One pair or family of each of these species foraged together day after day. At territorial boundaries, each species of one flock confronted individuals of the same species of the other flock in a noisy altercation that might continue for as long as half an hour, with much territorial calling, posturing, displays of usually concealed light patches, and chasing, but rarely a bodily clash. While a dispute lasted, flock members of a species not represented in the opposing flock continued to forage without becoming involved. Finally, the flocks withdrew into their own domains and calm was restored.

Flock members nested in their group's territory. While incubation was in progress, the off-duty member of a pair foraged peacefully with its usual companions, often more than 330 feet (100 m) from its nest. While a pair fed nestlings, they stayed closer to their nest, joining the flock when its daily rounds brought it within about 165 feet (50 m) of the brood to which they carried food. After the young could fly well enough, they accompanied parents who fed them regularly in the mixed flock. Even after they could forage for themselves, juveniles continued to travel with their parents in the flock.

In these mixed-species flocks of Amazonian Peru, the nuclear species, or leaders, were two antshrikes, the Bluish-slate and the Dusky-throated. Dull of plumage (except the bright red abdomen of the female Bluish-slate Antshrike), these flycatching antbirds were nervously active, frequently flicking their wings and changing their stance, and noisy, with loud, distinctive notes that they repeated frequently throughout the day. Their voices appeared to call the flock together as

Black-faced Antbird *Myrmoborus myotherinus* Male Eastern Colombia to Bolivia and Amazonian Brazil

it assembled each morning, and to guide its course through the forest. When one of these leaders moved ahead of the flock by as much as a hundred feet (30 m), its voice nearly always drew the others onward, perhaps with some delay. Of the two antshrikes, the Bluish-slate, which foraged higher and called with louder, more insistent notes, was the more effective leader.

The core of this flock consisted of one pair of each of six species who were equal shareholders of the group's territory. In addition to the two antshrikes, it included the White-flanked Antwren that we met in the Panamanian flocks and two related species, the Long-winged and the Gray antwrens. The sixth species was an ovenbird, the Rufous-tailed Foliage-gleaner. Ten other core members had sparser populations than the first six and usually larger territories, which enabled them to cross boundaries from one flock to another when two flocks met. Among these were two other antwrens, the White-eyed and Ihering's, the Plain Xenops that accompanied the Panamanian flocks of antwrens, several other ovenbirds, woodcreepers, and the Red-crowned Ant-Tanager. Another two antbirds, the Plain-throated Antwren and the Black-faced Antbird, holding separate territories smaller than that of the flock, joined and dropped out as the flock passed through their respective domains. Moreover, regular members of canopy flocks sometimes descended to accompany the understory flock led by the two antshrikes. With the occasional presence of other antbirds, woodcreepers, ovenbirds, flycatchers, wrens, tanagers, and even a little piculet, these mixed-species flocks could become very large, noisy, and conspicuous.

Although all these flock members were mainly or wholly insectivorous, as in the Panamanian flocks they reduced competition for food by hunting insects in different ways or at different heights. Thus, of the two flycatching antshrikes, the Bluish-slate commonly perched about twice as high as the Dusky-throated; and of the foliage-gleaning antwrens, the Gray foraged twice as high as the Long-winged. Like other forms of territoriality, the group territories of the mixed flocks promoted a uniform distribution of its diverse members. But perhaps the greatest benefit of participation in the flock was safety in a forest infested with at least six species of specialized bird-eating hawks. It is probably harder for a raptor to surprise a bird surrounded by many other pairs of watchful eyes than when it is alone. The two antshrikes that led the flock also guarded it, giving loud, unambiguous alarm calls on the approach of danger. Not only are animals of many kinds safer amid others than when solitary; it has been demonstrated, elsewhere than in South American forests, that they feed more securely and devote to foraging time that solitary birds might spend looking around for enemies, which is a substantial advantage of flocking.

At six o'clock in the morning or soon thereafter, the flock at Cocha Cashu assembled for the day's activities, apparently called together by the antshrikes. By seven the birds were quietly foraging, either in one place or while slowly drifting through the forest at an average rate of about two hundred feet (60 m) per hour. They continued to forage actively until noon, when many paused to rest or preen for intervals of about half an hour, while others continued to hunt food. In midafternoon they resumed more active foraging and continued until around five o'clock, when the flock scattered and many of its members bathed in a small woodland pool. After their ablutions they went to roost, apparently separately rather than all in the same tree.

The evening bath appears to be frequent among antbirds. After leaving a swarm of army ants as day ends, lone Bicolored Antbirds, or sometimes up to six together, bathe in forest rivulets, near or far from the ants they have been following. Sometimes Ocellated Antbirds join them in the water, not without frequent supplantings and some aggressive displays among the bathers. Black-spotted Bare-eyes also splash in the water as daylight fades. One drizzly October afternoon, I watched a female Dotted-winged Antwren wet her plumage in rainwater that had collected between the broad leaves of a tank bromeliad growing on a small tree at the edge of second-growth woods. After she emerged from the little aerial pool, one of the two males who accompanied her bathed in the same place, then all three flew back into the woodland.

Willis described how Spotted Antbirds take their evening bath. A pair chirped and sang faintly as they advanced slowly to a rivulet, looked down at several little pools, then dropped to the shallow margins of two of them. After hopping into the water about up to the middle of their legs, they sat and fluttered briefly. Then, half sitting with wings relaxed and tails spread, they paused and looked around before ducking their heads and foreparts and splashing with their wings, until drops cascaded down their backs when they rose. The female flew up and shook her plumage, then returned to her pool to immerse herself again. After the bath, the pair rested on branches of a fallen tree for a leisurely preening in the failing light.

Birds so strongly attached to the shade as antbirds have seldom been seen to sun themselves, and rarely as long as many other birds do. In a beam of bright sunshine, a Chestnut-backed Antbird fluffs out its body plumage, lifts its sunward wing to expose the underside to the rays, and tilts its head to gaze toward the sun with one open eye. For a minute or so it maintains this rigid pose, perhaps to dry its feathers or to kill lice or mites. A female Spotted Antbird was observed perched in a beam of sunshine with feathers fluffed and face turned toward the luminary for a similarly brief interval. A Bicolored Antbird who happened to alight on a sunlit perch reacted to the beam by momentarily assuming a sunbathing posture.

Rubbing ants or other items on the feathers—the widespread, puzzling activity known as "anting"—has been recorded by Willis of four army-ant followers at swarms: Bicolored, Spotted, Lunulated, and White-breasted antbirds, most frequently in the first of these. Holding the item in its bill, often mandibulating it before applying it to the feathers, the bird sweeps it over the underside of the primaries of an uplifted wing, or of the tail pushed forward between the legs. The item

Rufous-winged Antwren *Herpsilochmus rufimarginatus* Male Eastern Panama to northern Argentina

it uses may be an army ant, some other kind of ant, or a different small creature. After being rubbed on the plumage, the prey is either discarded or devoured; male Spotted Antbirds sometimes ate it and occasionally offered it to their mates. Willis thought that anting was most frequently practiced by hungry subordinate birds on the outskirts of swarms, who were reduced to picking up less desirable and possibly distasteful prey, from which they might rub some disagreeable secretion. Among the explanations adduced for anting are the destruction of external parasites by formic acid or other secretions, soothing skin irritated by molting, or producing pleasant sensations.

When not busy foraging, antbirds loaf and preen, not only themselves but often their mates, especially the heads and necks, where birds cannot groom themselves. One afternoon in the forest, I heard the harsh *waaa* of a Chestnut-backed Antbird, who was promptly answered by another on the other side of the path. The grating notes sounded back and forth several times before the more distant bird joined its mate. Then, while the two perched on adjoining twigs, one stretched up its head with all the feathers sticking out, leaving the skin of the neck largely exposed and giving the small bird a vulturine aspect. The other nibbled at the erected head feathers of the first. In the dim light, I could not tell which bird gave and which received this attention. I have also watched Spotted Antbirds preen their consorts. Antbirds often feed their partners, as will be told in more detail in the following chapter.

As day ends, antbirds seek their resting places, always in trees or shrubbery—never, as far as known, in dormitory nests or holes, as many other birds sleep. After leading a mixed-species flock all day, a pair of Dusky-throated Antshrikes roosted in low trees within 230 feet (70 m) of where the flock gathered the following morning. Other flock members went to different roosts. In Panama, the birds who associated during the day slept more socially. On a number of nights, a family of Dotted-winged Antwrens and a pair of Checker-throated Antwrens roosted in the same tree, with a Slaty Antshrike on the same branch with the latter. Around them within a radius of fifty feet (15 m) slept other flock members, including White-flanked Antwrens, Plain Xenops, Black-striped Woodcreepers, Wedge-billed Woodcreepers, and wintering Chestnut-sided Warblers. These birds who roosted close together could assemble and start on their foraging rounds by 6:30 in the morning.

Unless they occupy a hole or a dormitory nest that can be found in the daytime and watched at nightfall or dawn, sleeping birds of all kinds are extremely difficult to discover amid the heavy concealing vegetation of humid tropical forests. Of how antbirds pass their nights, our information is lamentably scanty. The best available study of an antbird's roosting was made by Allen M. Young, who in early February found by flashlight a male Spotted Antbird sleeping in the forest in northeastern Costa Rica. The bird was roosting about ten feet (3 m) up on a horizontal branch of a small, leafy tree. On many nights during the following four months, or until the end of May, he slept in the same tree, and on most of these nights on the same branch, where he arrived with great regularity, nearly always within a few minutes of six o'clock, whether the evening was clear or darkly overcast. Where he passed the night when he was absent from this tree, and where his mate slept, remained unknown.

4 *Voice, Displays, & Courtship*

With vocal organs simpler than those of the oscines or songbirds, antbirds appear incapable of the complex musical figures of many of the former, and they are not known to mimic. Nevertheless, the family as a whole has a great diversity of songs, and many species have various and contrasting notes for the different occasions of their lives. The voices of females are often similar to those of males but tend to be softer or higher in pitch. Mated birds frequently sing responsively or duet, sometimes while one of them sits in the nest. While singing or calling, antbirds often swing their tails up and down in time with their notes, and those with crests erect them.

A frequent song is a series of similar notes, often accelerated or rising in pitch, and ending with a contrasting sound. An example of this type of song is that of the Great Antshrike, a prolonged loud roll increasing in tempo, often becoming so rapid that it defies transcription: *took took too to to to t't' t'rrrrr.* This verse is often terminated by a nasal *waah,* a sort of buzzy growl of most peculiar intonation which does not carry so far as the preceding notes. Rather harsh when heard near its source, the powerful utterance is melodious when softened by distance. Raised crest, distended throat, and rapidly vibrating tail suggest that Great Antshrikes throw much energy into their songs.

Of the same type as that of the Great Antshrike but simpler, is the song of the much smaller Slaty Antshrike, which might be written *r-r-r-r-r-r-wánk.* Sometimes, especially at the nest, the order is reversed, the emphasized staccato note introducing the long, ascending rattle or roll: *wánk r-r-r-r-r-r.* Or the antshrike utters a simple *churr,* without an emphasized note. The Barred Antshrike's loud, dry rattle, sometimes continuing for four seconds, likewise terminates with a *wank*. With this rattle a mated pair duet, the two singing simultaneously, or the female following her partner with exactly the same strophe, although often in a slightly higher pitch. In addition to its wooden roll, the Barred Antshrike voices a crowlike *caw*, a guttural croak, and soft, mellow notes like those of the Black-throated Trogon. Its notes are sometimes confusingly similar to those of other birds and even those of frogs.

Softer and more appealing are the songs of the little Spotted Antbird. In addition to a high-pitched trill, it has at least two other songs, delivered in a voice that sounds far away even when the bird is near. One may be paraphrased *peede weede weede weede weede weede weede*, uttered slowly and slightly descending in pitch. The other song sounds much like *peede peede peede peede sip sip sip*. The female answers her partner with similar verses. At best subdued utterances, these songs are sometimes softened to whispers.

Pleasantly whimsical rather than musical, the Bicolored Antbird's song is a long sequence of clear, thin notes that are at first fairly long and distinct. As they are repeated with accelerated tempo, they be-

Spotted Antbird *Hylophylax naevioides* Male Eastern Honduras to western Ecuador

come shorter and rise in pitch, until at times they almost merge into a trill before they become longer and lower. The song may be represented thus:

```
                          _
                      _       _
                  _               _
              _                       _
          _
_   _   _   _
```

This song has many variants, one of which lacks the final falling cadence and may be written *wee wee wee wee wee wee wheer,* the last two notes rising conspicuously in pitch. Sometimes a harsh, nasal drawl, or several of them, terminate the performance. When foraging with army ants, these birds frequently voice, especially when disturbed by an intruding human, a drawled, complaining, throaty note. They also repeat a peculiar whining note and a low *churr.* With a harsh, rasping call, one Bicolored Antbird flies at and supplants another. At a well-attended ant swarm, with a number of these birds singing and calling on all sides in varying tones and cadences, the effect is most amusing and reveals that, although antbirds cannot sing like thrushes or mockingbirds, their voices are flexible enough to express a wide range of emotions and to communicate with their companions.

Contrasting with the churrs and wooden rolls of many antbirds are the more liquid whistles of others. Among the most characteristic sounds of Central American rain forests are the clear whistles that Black-faced Antthrushes repeat at intervals as they walk daintily over the ground. Usually an emphasized first note is promptly followed by two others—*whoo who who*—but sometimes by only one, and if excited, the antthrush may string together ten or more of these whistles. A skillful imitation is often rewarded by a view of the short-tailed bird approaching cautiously through the undergrowth. In the same forests, the Chestnut-backed Antbird's whistle is readily confused with that of the antthrush, but it is less bright and resonant, slightly more plaintive, more like a Black-throated Trogon's notes. Moreover, these whistles are heard in pairs or rarely triplets, whereas the antthrush commonly sounds three or more together. Mated Chestnut-backed Antbirds answer one another with similar whistles or with a harsh, nasal *waaa* or *aaaa*. They scold or express alarm with a rattle: *wittit wittit wittit wittit*. The male Dusky Antbird's soft whistles, rapidly repeated with ascending pitch, are answered by his mate with slightly higher notes in the same pattern.

Also heard in Central American rain forests, but more rarely, is a series of about seven to nine whistles not unlike those of the Chestnut-backed Antbird, but sounding more hollow, melancholy, and distant even when the source is near. These notes are repeated with increasing speed until about the middle of the series, then become more widely spaced as the song ends—*cow — cow-cow cow cow cow-cow — cow*—the whole sequence almost too rapid to count. Until I watched a Spectacled Antpitta singing on its nest, my efforts to identify this mysterious sound were always futile.

Soft twitters, chirps, cheeps, and low, whining notes are uttered by little antwrens. The Slaty Antwren's modest song runs *t'weet t'weet t'weet t'weet weet weet weet weet* in low, soft tones. The Dotted-winged Antwren's rarely heard song begins with six weak notes in ascending pitch, followed by a slight falling rattle: *che che che che che che chr'r'r'r'r*. When alarmed or perturbed, the Dotted-wing repeats, over and over, a plaintive *tweo*. The small Banded Antbird of northern South America sings with a prolonged, insectlike trill.

On the whole, antbirds sing or call rather sparingly, seldom as profusely as many thrushes and other songbirds sing at the height of their breeding season. An exception is the White-bellied Antbird, who on long legs hops over the ground tossing fallen leaves aside, or flits from twig to twig amid low, tangled vegetation. Through much of the day, it repeats, at intervals of a few minutes, a sequence of up to twenty-five loud, far-carrying, bold and clear but hardly musical notes. At first distinctly separated, the notes are accelerated into a trill that is followed by louder notes more deliberately uttered: *chee chee che che chechecheche che chip chu chu*. While singing the antbird stands on the ground or rests on a perch a foot or so high, pumping its brown tail up and down. Despite its attention-drawing voice, the White-bellied Antbird is not easy to glimpse amid low, tangled growth at the forest's edge or in light woods.

The most striking, frequently noticed displays of antbirds are those that expose the usually concealed patches of white or buff on their crowns, backs, or wings, as these birds do in boundary confrontations with neighbors of the same species, or when their nests or young are,

Dotted-winged Antwren *Microrhopias quixensis* Male (right) and female (left) Southeastern Mexico to Bolivia and central Brazil

or appear to be, endangered. Since nest defense will engage our attention in chapter 9, I shall here consider only displays in the former of these contexts. When two male Chestnut-backed Antbirds met in the undergrowth of the forest in southern Costa Rica, they perched not far apart near the ground, stretched up their necks, and at the forward edges of their wings, where most of the time no white was visible, displayed white patches that contrasted with the black of their under plumage. While posturing, they voiced their mellow double whistles and also an appealing note, between liquid and whining. Before I could see much of this altercation, the birds drifted away through the undergrowth.

More spectacular are the displays of Dotted-winged Antwrens. While I walked through the forest in the valley of El General on a morning in June, shrill, cheeping calls drew my attention to four of these antwrens, apparently two pairs, in the tops of taller saplings, ten to twenty feet (3 to 6 m) above the ground. Confronting each other, the females perched upright, with heads elevated and bills pointing skyward. The feathers of their backs and rumps were raised and spread outward, converting the whole upper surface into a fluffy puff, white on its forward face. Their slaty wings were relaxed, displaying

White-flanked Antwren *Myrmotherula axillaris* Male Honduras to Bolivia and southeastern Brazi

the conspicuous white areas on the coverts. Their tails were depressed and fanned out, exposing the prominent white feather tips. They appeared much bigger and more formidable than they actually were. When not flitting nervously from twig to twig, they rested face to face, repeating rapidly and almost incessantly a loud, sharp *peep,* interspersed with a grating, nasal rattle. They maneuvered back and forth, one often flying at her opponent, without ever coming to grips, as far as I saw. During the ten minutes that this game continued, the two males from time to time performed in the manner of the females, but they were less active participants in the encounter. Among the small onlookers of this noisy match was a male Black-hooded Antshrike, who rattled loudly.

In a later year, in the same locality, I watched at least five Dotted-winged Antwrens displaying together, males confronting males, females facing females, all showing the maximum amount of white in their dark plumage. As usual, this appeared to be a formal display of adornments rather than a fight.

As is frequent in birds, the same embellishments are conspicuous in both agonistic encounters and in courtship—they cannot change their clothes for different occasions. In May I saw a male Dotted-winged Antwren address a female, apparently his mate. With a patch of snowy white gleaming in the midst of his black upper plumage, he repeated sharp, clear notes so rapidly that they almost formed a trill. Before I could learn the outcome of this behavior, the chestnut-breasted female flew away. Covering his white dorsal patch, the male followed.

On a February afternoon on Barro Colorado Island, I found a male and two female White-flanked Antwrens flitting restlessly between low twigs, incessantly repeating a call of two slight notes. While so engaged, they lifted and closed their wings, one at a time, and very rapidly. As the tiny birds flipped up their wings, they turned from side to side, and about-faced on their perches. The male and females behaved in the same way, but the females called and exercised their wings more actively. When the black male raised a wing, he momentarily revealed the long white feathers on his flank. The females had no similar adornment to display. For about half an hour the birds continued these antics while they drifted through a small area of the underwood, neither coming close together nor performing face to face. In a subdued voice, one of the females churred rapidly. My impression that the female White-flanked Antwren may take the initiative in courtship was strengthened, two months later, when I watched a female pass to a male in adult plumage an insect that she had caught.

I have never again seen a female antbird give a morsel to her mate, and very rarely have I witnessed such a transfer of food in birds of other families. But among antbirds feeding of the female by her partner is widespread if not universal in the family. Among the male antbirds that I have seen pass food to their consorts on one or more occasions are the Black-hooded Antshrike, Dusky Antbird, Dotted-winged Antwren, White-flanked Antwren, Plain Antvireo, Spotted Antbird, Bare-crowned Antbird, and Bicolored Antbird. Others have recorded such feeding in the Slaty Antshrike, Black-crested Antshrike, Cinereous Antshrike, Dusky-throated Antshrike, Lunulated Antbird, White-throated Antbird, Ocellated Antbird, Hairy-crested Antbird, Bare-eyed Antbird, Harlequin Antbird, and Scaly-backed Antbird. Probably in most of these instances the male gave food to a female to whom he had long been mated, so that the activity was more properly nuptial rather than courtship feeding. As the date of laying approaches, males liberally nourish their consorts, helping them form the eggs. This does not exclude the possibility that gifts of food also help males win partners in the first place, as we have ample evidence that it does in a few species. Constant association and mutual grooming also help to preserve pair bonds that are fortified by nuptial

Hairy-crested Antbird *Rhegmatorhina melanosticta* Sexes similar
Southeastern Colombia to Bolivia and central Brazil

feeding. In Ecuador, I watched a Dotted-winged Antwren feed, then mount, his mate.

On Barro Colorado Island, Willis found male Bicolored Antbirds more numerous than females in a ratio of about 1.7 to 1. Almost twice as numerous as females, males often wander unmated. Timid young females whose parents have recently ceased to feed them forage on the outskirts of ant swarms, where they may not catch enough escaping creatures to assuage their hunger. High-ranking males forage so successfully that they become satiated and dally with the superfluous items they catch instead of promptly swallowing them. In this situation, a famished young female often solicits food from an overfed bachelor; she may even try to snatch it from his bill. If she persists in begging from him, the unmated male may reluctantly indulge her. At first his behavior toward her is ambivalent; after she has received his insect he may drive her away with a hiss and a snap. As days pass, he becomes more tolerant, permits her to forage near him, and often passes food to her. If this continues for several weeks, the pair bond is established; the pair travel together from ant swarm to ant swarm, and in due season seek a nest site.

Instead of grunting at one another, as in the early days of their association, the newly wedded Bicolored Antbirds converse with chirps and faint songs. When separated, they answer each other with louder voices. As they pass from one ant swarm to another, the male leads the way with song and flicking tail; they rarely stay with one swarm for more than a day or two. Living as equals, neither tries to dominate the other nor assumes a submissive attitude toward his or her partner. During a honeymoon that may continue for several months, one rarely tries to supplant the other on a perch.

This idyllic state does not continue indefinitely. After they have been long together, the partners grunt at and supplant one another more often, although mated birds rarely challenge each other strongly or cringe deeply. Usually the female supplants her partner, who flutters unresistingly from his perch. If he delays there after she has alighted beside him, flitting his wings, chirping, or gaping, she sends him flying with hisses and snaps. Now the mates forage a yard or two apart. Of eighteen males closely watched by Willis, thirteen were subordinate to their consorts, only five were dominant members of a pair. Their status appeared to depend upon individual character or temperament; a male who dominated his first mate was subordinate to a later consort, the daughter of a henpecked male. Since at first, when young females beg for food from their future husbands, they are necessarily subordinate to them, how they eventually win ascendancy is puzzling.

Although at any season a male Bicolored Antbird may pass to his partner morsels that he is too full to swallow, as the date of nesting approaches he feeds her more frequently, even before he has satisfied himself. Soon he is giving her so much that she ceases foraging for herself and spends her time loafing and preening in dense cover near the swarm where he catches insects for her. If he relaxes his effort to nourish her, she may become aggressively importunate, lunging so rudely when he alights beside her with a gift of food that she bowls him backward from the perch. As days pass and she feels more dependent upon her servitor, she becomes less imperious. After giving her a cockroach or a juicy spider, the male may hop upon her back while she struggles to swallow it. Not yet ready for this, she flutters away. A few days later, she may permit her mate to mount her while, appearing more interested in food than in sex, she chews what he has just passed to her. Henceforth, the female finds more food for herself and receives less from her partner; coition becomes less frequent as eggs are laid and incubation begins.

Not every intimacy between a young female and an older male who caters to her ripens to marriage. A female recently cast off by her parents may be fed—for a day, a week, or more—by a succession of males during her first months of independence. Young, subordinate males occasionally feed young females at swarms where competition for food is not severe; but only older, more capable males are likely to procure enough to retain a female long enough to mature a nuptial bond. Males rarely win mates when less than a year old; some perforce remain bachelors for five or six years, or all their lives. Females often pair when four to six months of age. Matrimony and maturity do not end their juvenile propensity to snatch food from males to whom they are not mated. Such dalliance may continue until a flirtatious female's spouse arrives. However, in the case of a female who has been widowed, this may be the beginning of a new alliance.

Sometimes two mateless male Bicolored Antbirds become close companions, almost like a male-female pair. They visit ant swarms together, feed each other, and pass an insect back and forth between them. Some are widowers who have already reared offspring. A male who had lost his mate formed a close friendship with another male, several years old, who had never succeeded in winning a female. Some of these males may eventually find consorts and nest. Rarely a mateless male feeds a very young and meek male as though he mistook him for a female.

Spotted Antbirds always form pairs while one member is on its territory. A male waiting for his first partner, or a widower, sings loudly and persistently, while he flirts with and feeds any passing female. If this continues for a few days, the two remain united. A female less than six months old may accept a male's gift of food without joining him in matrimony; when courted, she behaves like a juvenile. Most new spouses are females from six to twelve months of age, or older females who have wandered from their territories after losing their mates. Like Bicolored Antbirds, mated female Spotted Antbirds sometimes accept food from males other than their partners; and mated males are not always above courting females who are not their consorts. Such flirtations may disrupt pairs. Once two males exchanged mates, but polygyny apparently never occurs among Spotted Antbirds.

Chestnut-belted Gnateater *Conopophaga aurita* Male Northern South America

A male Spotted Antbird feeds his partner generously for a few days before he attempts to inseminate her. At first she hops away when he tries to mount her, but little by little her reluctance is overcome, and she invites coition while she mandibulates the insect that he has just given to her. More closely united than Bicoloreds, mated Spotteds rest almost in contact while they preen themselves or each other, whereas the former preen more than their body width apart, and supplant mates who perch less than a yard away. So close is the companionship of mated Spotted Antbirds that it is difficult to tell which sex, if either, dominates the other.

Edwin Willis's thousands of hours of watching birds at ant swarms, often in fairly open situations, gave him exceptional opportunities to observe the interactions of his color-banded birds, including the formation of pairs. Little is known about pair formation in antbirds that rarely follow ants but flit obscurely through dense thickets or forest verdure. Indeed, in tropical birds of many families that live in pairs throughout the year it is far more difficult to learn how pairs originate than it is to follow the courtship of migratory birds that mate anew at the outset of each breeding season.

Quite different from the display recorded of any more typical antbird is that of the male Rufous Gnateater. In the evening twilight of the breeding season, he flies around and around his territory, making a loud rattling noise with his wings, on which the outer primary feathers are expanded at the tips. He also sings a simple whistling melody ending with a few lower emphasized notes. The female's wings are of normal shape and her flight is silent.

5 *Nests*

Many antbirds build open cups, attached by the rim to the arms of a horizontal fork of a slender branch or in a similar situation. The size and composition of these vireolike nests vary greatly. A bulky nest of a pair of Great Antshrikes was suspended between two slender, diverging stems of the scrambling composite *Eupatorium vitalbae.* Long, slender, dry herbaceous vines, looped over the supporting arms, formed a wide-meshed basket that held the thick layer of leaves which composed the bulk of the structure. Most were pieces of such great-leaved monocotyledons as *Heliconia* and *Calathea.* Some of these strips, broad and long, were coiled and twisted into the nest. Thin, curled, dry herbaceous vines lined the bottom of the open cup, which was 6 inches (15 cm) high by 5 inches (12.5 cm) in overall diameter and nearly 5 inches deep.

Rather similar in construction but less bulky is the Russet Antshrikes' nest. A thick layer of coarse, dry leaves, in this case of dicotyledonous trees, is held in place by thin, dark filaments—fine rootlets or the fungal rhizomorphs often called "vegetable horsehair"—looped over the supporting arms and intercalated with the leaves. In one nest an inner layer of finer bamboo leaves was overlaid by a few dark strands as a lining. Attached to the outside was a spray of a creeping polypody fern with tiny, roundish green leaves. These nests blend well with the foliage of forest trees amid which they are usually situated, up to 50 feet (15 m) above the ground, the highest antbirds' nests that have come to my attention.

The nests of *Thamnophilus* antshrikes that I have seen—Slaty, Black-hooded, Barred, and an unidentified species in Peru—differed from the foregoing in their much thinner walls, without leaves. In most of these nests the fabric of fine, mostly dark fibers was so open that light passed through, or the eggs could be glimpsed from below. More or less green moss decorated the exterior, but a pair of Barred Antshrikes had substituted for the moss a few sprigs of the delicate inflorescence of a weedy *Iresine.* Of similar construction, but larger, was a Fasciated Antshrikes' nest, 25 feet (7.7 m) above a Costa Rican stream flowing between high, brush-covered banks. Nests of Plain and Streaked-crowned antvireos are also open fabrics of dark fibers, decorated with strands of green moss that sometimes dangle beneath them. All these nests are only about half as large as the Great Antshrikes' capacious structure.

Antwrens' nests vary more in composition. Those of the Slaty Antwren that I have seen were deep open cups, fastened to the arms of a horizontal fork or to two nearly parallel branchlets, with the body of the nest hanging below these supports. Blackish fungal strands and fine, dark-colored rootlets formed a fabric so open that the eggs were visible through the bottoms and walls of the delicate pensile structures, devoid of moss. Similar in form and attachment, nests of White-flanked Antwrens are composed mainly of pieces of dead leaves or their lacy skeletons, bound together and fastened to the supports by vegetable horsehair, more of which thinly lines the bottom. The Dotted-winged Antwrens' nest is also made of dead leaves held together by dark strands. Similar in form, the White-fringed Antwrens' flimsy nest is composed of fine grasses. The Wing-banded Antbirds' shallower cup of twigs and rootlets in a horizontal crotch resembles a manakin's nest.

Instead of being attached to a horizontal twig, the Checker-throated Antwrens' nest is fastened in a fork near the end of a slender, drooping branch in the undergrowth of the forest. A purse rather than a cup, it has an oblique opening at the top, between the branchlets that support it; it may measure up to 6 inches (15 cm) long by slightly more than 3 inches (7.5 cm) wide near the bottom. Fine black fibers, rootlets, partly decayed leaves, and bits of herbaceous stems compose the wall of dead leaves; fine fibers line the bottom.

The Dusky Antbirds' pouch also hangs between two divisions of a drooping branch, vine, or fern frond. The deep pocket is widest near the bottom and narrows to the orifice, which is so strongly oblique, facing upward and outward, that it makes the structure much higher at the back than at the front. Nests that I measured were 5 to 7½ inches

Nest and eggs of Slaty Antshrike *Thamnophilus punctatus*

Nest of Checker-throated Antwren *Myrmotherula fulviventris*

(12.5 to 19 cm) in overall length by 3 to 4 inches (7.5 to 10 cm) in greatest diameter. The thick walls and floor were composed largely of dry dicotyledonous leaves, papery strips of monocotyledonous leaves, grass blades, or fragments of fern fronds, all loosely bound together and attached to the supporting fork by black fungal strands, brown fibers, and the like. A few fibers were coiled into a flattish mat on the bottom. Some Dusky Antbirds' nests have green moss about the rim and on the outside. Cobweb helps to bind the smaller antbirds' nests together and to their supports.

Not all antbirds' nests are attached by the rim, like those already described. Among those that rest upon the bottom are the bulky open cups or bowls of Chestnut-backed Antbirds and related species, loosely set near or on the ground in deep forest shade, in or beside low, spiny palms, on fallen branches or palm fronds, or amid concealing herbs. A foundation and outer frame of fern rhizomes, rootlets, sticks, lengths of decaying vines, inflorescences, or similar coarse materials supports and surrounds a thick middle layer of assorted leaves and fragments of palm fronds up to 8 inches (20 cm) long. The many long, coarse strands of vegetable horsehair and fibrous rootlets that line the concavity fail to conceal the leaves on which they are coiled.

Two reported nests of White-plumed Antbirds were open cups sunk, about a foot (30 cm) above the ground, in a mass of dead leaves and stalks in the spiny crowns of small palms amid the forest. Composed largely of dead leaves, these nests were lined with dark fibrous rootlets. Rather similar but situated somewhat higher are the bowls of Cinereous Antshrikes. Composed of dried decaying leaves and rhizomorphs, they look like piles of debris. Likewise, a loose accumulation of dead leaves and twigs surrounds gnateaters' compact cups of fibers and makes them hard to detect. Unlike the great majority of antbirds' nests, that of the White-shouldered Fire-eye is a roofed structure with a side entrance, like the Great Kiskadee's nest. Made of straws and dry strips of leaves, especially of plants of the arrowroot family (Marantaceae), it rests, on or near the ground, on low tree stumps or amid ferns, as described long ago by C. Euler in Brazil.

Antpittas are not the most skillful of nest-builders. Better made than most antpittas' nests is that of the Thrushlike Antpitta. Two nests of this little bird, found in French Guiana, were thick-walled cups of twiglets, lined with rootlets. They were built, about a foot above the ground, on beds of dead leaves between the fronds of a terrestrial fern and on a clump of herbaceous plants. The Rusty-breasted Antpitta is also a fairly competent builder. On a small bush or tangle of vines, 2 to 4 feet (0.6 to 1.2 m) above the ground, the stubby-tailed bird builds a platform of coarse sticks, upon which it fashions a shallow cup of curving rachises of the compound leaves of a mimosalike plant, all placed with their curvatures inward. It adds no other lining.

Nests of the Spectacled Antpitta are untidy, loosely made, shallowly concave platforms of coarse twigs, dry petioles, and large dead leaves, often precariously situated from 2 to 5 feet (0.6 to 1.5 m) up in a shrub, tree fern, or epiphyte beneath tall forest trees. Much of the material appears to have fallen on the nest site before the antpittas added a scanty lining of dark fibrous rootlets. Occasionally, they line the abandoned open nest of some more competent builder. Resting upon fallen logs or branches amid tangled undergrowth, nests of the Scaled Antpitta are shallow bowls composed of dead leaves, pine needles, small sticks, and moss, lined with fibrous rootlets. In Panama, a big Black-crowned Antpitta built a similar nest on a bed of dead leaves in the low crown of a tagua palm.

A few antbirds lay their eggs in enclosed spaces. The usual site of a Bicolored Antbirds' nest is a low stub of a palm trunk, or less often of a small dicotyledonous tree, reduced by decay to a fragile hollow shell opening skyward. These stubs are from a few inches to, rarely, 5 feet (0.1 to 1.5 m) high, with cavities 2½ to 4 inches (6 to 10 cm) in diameter. If the hollow is deep, it is partly filled with strips of dry palm fronds mixed with fragments of dicotyledonous leaves, above which the builders arrange a thin mat of rootlets and other fibrous materials. On this mat the eggs rest, 2½ to 4½ inches (6 to 11.5 cm) below the lowest part of the often oblique or irregular rim of the hollow. Less often the nest is built in the tubular sheathing base of a large, fallen palm frond that has lodged more or less upright in the undergrowth of the forest, or it is situated between the trunk of a palm tree and the still-attached stub of the petiole of a fallen frond. Nests of the closely related White-cheeked and Rufous-throated antbirds are similarly situated. Black-faced Antthrushes rear their families in much deeper hollows.

On a day in May, while I sat in a blind amid dense forest undergrowth, watching an incubating manakin, I noticed an empty nest, a neat cup of black fibers attached to the low fork of a sapling, in view of my left window. In the course of a day, it was visited by two antbirds of different species. First a male Slaty Antshrike, passing with his mate, paused to perch on the rim and voice a low, not unmusical churr. Later, when a pair of Spotted Antbirds came by, the male alighted beside the nest and whispered an appealing little song, then continued onward with his partner and did not return. Neither of the females gave any attention to the nest, which I could not identify because I never found it occupied. At another time, when a swarm of army ants with attendant small birds passed by an unfinished Spotted Antbirds' nest, a Bicolored Antbird, of undetermined sex, alighted on its rim. These episodes reveal the interest that antbirds, usually males, take in nests, even those quite different from their own, especially early in the breeding season.

Birds' visits to alien nests appear to be responsible for some of the erroneous attributions of nests in the literature of Neotropical ornithology. The nests of many birds cannot be identified with certainty without prolonged watching from concealment. Since bird collectors, eager to increase the number of their specimens, may shoot the first bird they see at a nest instead of patiently watching it, their records are not always reliable.

Edwin Willis told how, after a male Bicolored Antbird has been feeding his partner for several weeks, they search for a nest site. Leaving a swarm of army ants, often after it folds, the pair wanders through the forest, exploring. When the male finds a hollow stub or some similar cavity, he "nest shows," an activity widespread among birds that breed in holes but also practiced by some with open nests. Alighting on the top of a promising stub, he peers into the hollow with one eye and then the other, while he flicks his partly fanned tail and sings. Becoming bolder, he drops into the dim cavity, to rebound with a dead leaf or fragment of wood that he tosses into the air. At intervals he looks around to assure himself that his partner is watching. She may be visiting potential nest sites on her own initiative or inspecting some that he has already shown her; sometimes she begins the quest of a site and he follows. Occasionally, the male nest shows at a site that is quite inadequate. When they find a promising nook, one or both partners may enter to toss out debris or peck or tear projections from the rim.

At a Russet Antshrikes' nest at the edge of high forest beside a clearing, I enjoyed an exceptionally favorable opportunity to see how a vireolike nest is started. The site chosen by the pair was the roughly triangular space between a slender horizontal twig of a tall *Goethalsia meiantha* tree and two curving lateral twiglets that met a few inches out from their points of attachment. Screened above by foliage and quite inaccessible by humans, it was clearly visible from the open pasture. While I stood watching early on an April morning, the antshrikes gathered cobweb or cocoon silk and wrapped this material around the twigs that enclosed the space chosen for the nest. In three quarters of an hour the two brought material twelve times, and on three occasions they were at the nest site together. I could not distinguish their sexes.

By 6:30 next morning, they had fastened fragments of dead leaves to a strand of cobweb that crossed the open space. The birds continued to bring materials until, by 7:00, the nest space was loosely covered with pieces of leaves. Then, for the first time, I saw one of the builders sit in this space and, spreading its wings over the twigs at the side, try to shape the structure. However, the antshrikes continued to add and arrange most of their contributions while they rested on the supporting twigs. Repeatedly, the loose fabric over the nest space broke away, to hang precariously from one side. The birds spent much time pulling up the weft of cobweb and leaf fragments, spreading it over the nest space, and fastening it more securely to the enclosing twigs. They also brought some long, thin vegetable strands and carefully worked them into the fabric to strengthen it. Their procedure reminded me of that of a pair of Yellow-winged Vireos who built a nest of similar form, but with pieces of green rather than of dry leaves and at a faster pace.

Building went slowly. In the first hour of my watch, the antshrikes made eighteen visits to their nest, and in the second hour only ten visits. On arriving, the builders alighted near the base of the slender supporting twig and hopped rapidly outward to the nest. On most visits, they spent considerable time arranging its materials. To leave, they flew directly from beside the nest; usually one went as the other arrived. Sometimes they were together at the nest for a few seconds; but as far as I could see, each nearly always arranged what it had brought; only once did one appear to pass its contribution to its partner. When birds of other kinds find their partner sitting in the nest when they arrive with materials, they may pass them to it for arrangement, but this is not the way of antbirds, who prefer to work into the nest what they have themselves brought to it. When I left after watching for two hours, the central space was still open, with a rim of materials around it. To start this nest was not easy.

On the third day of building, the antshrikes worked faster, bringing twenty-three contributions in the hour from 6:30 to 7:30 A.M. After this, their work languished for five days, until a good afternoon shower ended a fortnight of unseasonably dry weather. By the following morning, the builders had at last completely closed the bottom and sides of their cup. Probably their material had proved too refractory when dry, and like other birds the antshrikes preferred to build with damp pieces that are more easily molded into the desired shape. In the half hour from 7:00 to 7:30, the pair added sixteen billsful of material, after which they stayed away so long that I grew tired of waiting for them. However, the structure was soon finished, nine or ten days after it was started. The rather deep cup was brown from the fragments of dead leaves that composed most of its bulk. A few large pieces of dry leaves hung loosely from the outside.

In mid-February, a pair of Dusky Antbirds began a nest low at the forest's edge, beside a banana plantation in Panama. When I first saw it, a flimsy weft of fibrous materials with a few dead leaves outlined what was to become a deep pouch of the form already described. While I watched a few feet away in plain view, the male and female flew up and each added a billful of material. Early next morning, I watched from a blind. The blackish male was the first of the pair to appear. Emerging from the bushes, he alighted on one of the arms of the supporting fork and twittered softly. A moment later, his brownish mate arrived, stood on the other arm of the fork, facing him, and joined him in twittering. Then both flew back into the bushes at the woodland's edge, to return after a few seconds and repeat this charming act. The male was the first to begin work; after coming three or four times to perch on the nest with empty bill, he brought a dry tendril from a vine tangle and deposited it in the bottom of the pouch. After his departure, the female came again with empty bill and arranged more carefully the tendril he had brought.

The male was more active in carrying materials to the nest. In two hours, he brought twenty-nine contributions, she, only seventeen. Frequently, she came empty-billed and rested on a supporting twig while she tucked in loose ends of fibers and tidied up the structure, or she sat snugly in the growing pouch and shaped it with her body and feet.

Spotted-crowned Antvireo *Dysithamnus puncticeps* Male Costa Rica to western Ecuador

The male also attended in no small measure to these details, and he always put in place whatever he brought. But it appeared to be the female's special concern continually to inspect the work as it progressed and to give the structure its proper shape. Nevertheless, he appeared to be the mainspring of the undertaking, for after intermissions when the pair foraged beyond my view, he was usually the first to return to the nest. Then he called his coworker with fine twitters, which grew into a louder trill if she delayed long. As soon as she joined him, they resumed building, bringing fine and coarse fibers, broad blades of dry grass, and strips of palm fronds. These were added to the nest in no definite order, with the result that the growing fabric was a random mixture of all these components. When the birds sat in the nest to shape it, using their toes to entangle the fibers, a foot sometimes broke through the still-frail bottom. Twice, while sitting in the nest, the male sang his cozy little trill. This nest was not finished, however, probably because the Dusky Antbirds had chosen an unstable support.

While building their dangling pouch, a pair of Checker-throated Antwrens also perched side by side on its top before starting a spell of work. The short, broad pieces of brown, dry palm fronds or similar leafy materials that they brought for the middle layer of their pouch were large for such tiny, short-tailed birds, and of much the same color. Although the male had a black throat spotted with white that the female lacked, in the dim light of the forest undergrowth I could not always distinguish their sexes. Each partner arranged its own contributions, pressing itself down until nearly hidden in the pouch as it molded the piece to the nest's contour. Even when one antwren alighted on the rim while the other was within, it did not pass its piece to the other but waited until it emerged. While resting near the nest or sitting in it, these antwrens often sang, rapidly repeating a soft *cheep*.

Each time a pair of Dotted-winged Antwrens returned to resume work on their nest, they arrived, softly twittering, with empty bills. The male went first to perch on the rim and called his mate with more twitters and peeps. She followed and entered to shape the pouch, which was so deep that only her bill and the tip of her tail were visible above the edge. After she emerged, the male sat briefly in the nest.

TABLE 1. *Participation of the Sexes in Building by Antbirds*

Species	Hours watched	Visits to nest: Male	Visits to nest: Female	Locality	Source
Black-crested Antshrike	2.5	28	6	Suriname	Haverschmidt 1953
Plain Antvireo	1	8	3	Costa Rica	Skutch 1969a
Dotted-winged Antwren	2.5	9	8	Panama	Skutch 1969a
Dusky Antbird	2	29	17	Panama	Skutch 1969a
Spotted Antbird	2	19	13	Panama	Skutch 1946
Spotted Antbird	1.9	21	10	Panama	Willis 1972a
Totals	11.9	114	57		

Then they searched among surrounding trees and vines until each found a small dead leaf and carried it to the nest. In their first bout of building, each added five dry leaves or leaf fragments to the structure. They worked in exactly the same way, each placing its own contribution to the nest, never passing it to the other for incorporation. Three more spells of work were even shorter; in two and a half hours, the male brought material only nine times; the female, eight times.

Although the nesting of only a few of the 250 species of antbirds has been studied, these include enough diverse types to suggest that the participation of both sexes in building is widespread, if not universal, in the family. Often the male takes the lead and does a little more than his consort (table 1). With the exception of a few species, such as the Slaty Antshrike in forest and the Barred Antshrike in semi-open country, antbirds' nests are notoriously hard to find, and opportunities to watch sustained building are rare. Antbirds that travel with mixed-species flocks may build only briefly while their flock passes near their nest. After a disappointing vigil in the early morning, when finches, tanagers, and other songbirds build most actively, one may find a pair of antbirds busily at work later in the day.

A pair of Plain Antvireos built their open cup in less than two days. A pair of Black-hooded Antshrikes took five or six days to finish their more ample cup; and a pair of Checker-throated Antwrens needed more than eight days to construct a deep pouch.

6 *Eggs & Incubation*

After a nest is finished, several days elapse before the female lays her first egg. The second is deposited two days later. Throughout the family, the usual number of eggs is two. Sets of three are exceptional in nests of Great, Barred, Slaty, and Rufous-winged antshrikes. In arid southwestern Ecuador, S. Marchant found three eggs in six nests of Collared Antshrikes and two eggs in four nests. Occasionally, Chestnut-backed Antbirds and others are found incubating single eggs, but not often enough to prove that this number is frequent in any species or population. Although Plain Antvireos nearly always lay two eggs, in Trinidad a female more than eight years old incubated solitary eggs in two successive nests watched by Alan Lill and Richard ffrench.

Antbirds' eggs are white, cream, buff, pale gray, greenish, or blue, usually more or less heavily spotted, streaked, blotched, or otherwise marked with shades of brown, reddish brown, or lilac. The Chestnut-backed Antbird's beautiful eggs are heavily blotched, speckled, and streaked with deep, rich purplish or rufous brown on a dull white ground. Heaviest on the egg's thick end, where they almost obscure the ground color, the markings run in streaks down the sides of the egg and thin out toward its sharper end. Hardly less attractive are the Bicolored Antbird's eggs, adorned with reddish brown or cherry-color on a whitish or creamy ground. As in many hole-nesting birds, eggs of the Black-faced Antthrush are unmarked white, but soon after they are laid they become heavily soiled, apparently with earth from the bird's feet, and stained brown by contact with the decaying vegetable matter on which they rest.

On the short, blunt eggs of the Spectacled Antpitta, deep brown blotches sometimes cover more than half of the bluish gray or brown-

Great Antshrike *Taraba major* Male Southeastern Mexico to northwestern Argentina

ish gray ground color. The Black-capped Antpitta's eggs are pale buff marked with chocolate or chocolate-brown. The eggs of other antpittas run toward greenish and blue. Those of the little Rusty-breasted Antpitta range from very pale emerald to grayish green or pearl gray, mottled with reddish and darker brown. The Thrushlike Antpitta's eggs are blue-green to turquoise-blue, marked with brown, but the robin-egg blue or turquoise-blue eggs of Scaled Antpittas and Variegated Antpittas are unmarked.

In all antbirds that have been watched at the nest, the female incubates at night and the sexes sit alternately by day. Both sexes develop bare incubation patches on the abdomen for more efficient transfer of heat to the eggs. I passed the whole of one morning and most of an afternoon in a blind before the nest of a pair of Great Antshrikes. Around sunrise, a prolonged, throaty, rolling call, beginning loudly and gradually dying away, announced the male's approach through the thicket. Since my arrival in the early dawn, the female had been sitting so low in the deep basket that I did not see her until he aroused her by alighting on the rim, whereupon she left and he took her place. For nearly two and a half hours, he sat in silence. When, at 8:24, his partner returned and stood above him on the nest's edge, he appeared reluctant to go. When, finally, he made way for her, she looked carefully around before she settled on the eggs. She remained sitting until, at 12:23 P.M., she ended her four-hour session when, with a rapid, harsh, clicking call, the male alighted on the rim above her.

On the second following afternoon, during a lull in the hard rain that had begun before midday, I found the female sitting when I resumed my vigil at 1:25. Soon I heard the male's long, accelerated call, sounding louder with each repetition as he approached through the thicket. At 1:44 he emerged from the bushes and hopped upon the nest above his mate, who promptly departed. After inspecting the eggs with his head deep in the cup, he jumped in to warm them. For the remainder of the afternoon, he continued quietly to cover them beneath a steady rain. At 5:42 P.M., his consort arrived with a long-drawn rattle and quickly replaced him. She was still sitting when I left her in the rainy dusk. Except for two brief interruptions when the female was disturbed, the eggs had been incubated continuously while I watched. The male's sessions of nearly two and a half hours in the early morning and four hours in the afternoon were separated by the female's four-hour turn on the eggs (table 2).

At this nest, as at that which I found nearby in the following year, both parents sat very closely. They clung to their eggs until I approached to within a foot or two, then dropped into the low, dense vegetation around the nest and raised peculiar, loud, protesting outcries. Sometimes the male sat until I shook the nest; then, if I disappeared into the blind, he would return to his eggs in two or three minutes.

The pattern of two long diurnal sessions by the male, separated by one by the female, suggested by my observations at the Great Antshrikes' nest, was confirmed by a much longer study of incubation by the very different Spectacled Antpitta. A flimsy nest three feet (90 cm) up in the leafy crown of a tree fern beside a woodland path was attended by a pair of which one member had a small patch of white feathers on the back, enabling me to distinguish the partners in a species whose sexes are nearly always quite alike. Since the normal member incubated by night, I concluded that her mate with a touch of albinism was the male. In the dim light of early dawn she usually left her eggs, which remained exposed until her partner arrived from a few minutes to nearly an hour later. One morning, however, she prolonged her nocturnal session until he replaced her at 6:46, well after sunrise.

TABLE 2. *Diurnal Incubation by Antbirds*

Species	Hours watched	Sessions (in minutes)			Intervals of neglect (in minutes)			Attendance[1] (percent)
		Number	Range	Total	Number	Range	Total	
Great Antshrike								
male	11	2	146–238	384	1	17	17	97
female		1	239	239				
Slaty Antshrike								
male	6	3	33–142	219	2	11+	22	94
female		2	53–66	119				
Plain Antvireo								
male	15	5	103–137	586	8	1–18	45	95
female		4	41–79	261				
White-flanked Antwren								
male	12.5	4	3–174+	333	5	11–94	178	76
female		2	98–140	238				
White-flanked Antwren								
male	6	1	134	134	2	75–90	165	50
female		1	33+	33+				
Dusky Antbird								
male	14	7	11–153	401	7	1–59+	154	82
female		3	79–124	291				
Chestnut-backed Antbird								
male	11	5	45–95	344	5	1–39	90	87
female		2	92–136	228				
Spotted Antbird								
male	12	3	44–217	338	4	8–22	54	92.5
female		3	36–164	327				

[1] Percent of time between end and beginning of female's nocturnal sessions that eggs were incubated. The plus sign indicates that the bird continued to incubate, or that the nest remained unattended, after the watch was terminated.

The four of his morning sessions that I timed lasted, respectively, for 5 hours and 10 minutes, 6 hours and 29 minutes, 6 hours and 4 minutes, and 6 hours and 13 minutes. Around midday, his consort replaced him, to sit for two or three hours, after which he returned to take charge of the nest until, after another two or three hours, she came in the late afternoon to begin her long nocturnal session. This incubation schedule is similar to that of the Resplendent Quetzal.

On the few occasions when I saw an antpitta approaching its nest for a turn of incubation, it came hopping over the ground, sometimes tossing aside fallen leaves. More often I failed to notice the bird until it flew up to the nest. Then, if its mate were still present, the latter promptly left; the changeover was always made silently, without ceremony. The birds always incubated in silence, except one morning when the male stretched up his neck and, his white throat prominent, for nine minutes repeated at short intervals a subdued version of the *cow cow cow* song, while faint answers came from the distance. Sometimes, before leaving the nest in the dim early light, the female sang softly in much the same manner. These birds sat motionless for long intervals, only rarely rotating to face in a different direction. With neck retracted, they appeared very rotund. The light rim around each dark eye made them appear alert. At intervals they closed their eyes halfway or a little more, never fully or for more than a few seconds. Not shy, they sometimes continued to sit while I walked along the path beside the nest, as they were more likely to do if I refrained from looking directly at them as I passed about two yards away. They were not perturbed while for long intervals I watched them, sitting unconcealed in the path about twenty-five feet (8 m) from their nest.

Long sessions of incubation are also practiced by Bicolored Antbirds. At one nest, I watched the male sit during forenoons for four to nearly six hours continuously. Then, instead of waiting to be relieved by his mate, he went off to seek her at an ant swarm and send her for her turn on the eggs. In the afternoons the two parents incubated alternately for shorter periods. These birds were also strongly attached to their nest in a low, hollow palm stub that opened upward. They would continue to sit in it, gazing up at my face, while I bent over it to peer down at them. If I made them jump out, they returned in a few minutes, while I watched only a few yards away. Nevertheless, these parents were inconstant in their attendance and left their eggs exposed for long intervals, once for an hour and a half, once for thirty-seven

Spectacled Antpitta *Hylopezus perspicillatus* Sexes similar Northeastern Honduras to western Ecuador

minutes while rain poured into their nest. Sometimes they brought a rootlet to add to the lining beneath their eggs, and dived headfirst into the narrow hollow.

Other antbirds change over more often and incubate for shorter intervals. Nevertheless, some of their sessions on the eggs are surprisingly long for such small birds. A male Plain Antvireo sat for well over two hours continuously, his mate for over an hour. One male continued calmly to incubate while I set my camera on a tripod, focusing down on his low nest from the distance of a yard and, standing unconcealed by the camera, made four time exposures. While I watched another low nest, a black tayra—a large, highly predatory weasel—walked between it and my blind. When about three yards from me, the mustelid stopped and looked suspiciously in my direction, sniffing the air, then resumed his walk at the same deliberate pace. The female antvireo, who was then sitting, remained at her post on the eggs.

Dusky Antbirds have also continued to incubate while I set up the blind from which I intended to watch them, a short way off, and they have remained in their hanging pouches until I shook them. A male took sessions up to two and a half hours long, and his mate once sat for two hours. Whenever one partner sat until the other came, they exchanged places in the same manner. The new arrival alighted on the slanting lateral rim of the deep pouch, above its sitting mate, who after a few moments emerged and flew away. Then the newcomer slowly and carefully lowered itself into the pouch, as though it were a tight fit, somewhat as one might insert an oversized cork into the mouth of a bottle. Both sexes always sat with their heads at the front of the nest, where the rim was lowest, their tails held erect against the high wall at the rear. When the incubating bird was perfectly at ease, its head was just visible at the front of the pouch. When the male sang in the distance, his incubating partner answered with trills weaker than his. Like Plain Antvireos and Bicolored Antbirds, Dusky Antbirds occasionally bring a fiber when they come to incubate.

Chestnut-backed Antbirds approach their nest, low amid forest undergrowth, by hopping with pumping tail over the ground and along the log or fallen branch that supports their open bowl. In the early morning, a male repeated loud double whistles as he advanced, but through most of the day both sexes voiced a loud, harsh *aaaa* as

Plain Antvireo *Dysithamnus mentalis* Male incubating in southern Pacific Costa Rica

they neared their nest. The sitting partner, especially the male, might answer with a similar but less loud call. As he left, he might continue these grating sounds; but the departing female uttered low, soft notes or else left silently. Although she took longer sessions than he did, up to an hour and a quarter, he took more of them, and during the period of diurnal activity the male was on the nest 50 percent of the time. Together, they kept the eggs covered for 87 percent of the day (table 2).

The most prolonged study of incubation by Chestnut-backed Antbirds was made by Yoshika Oniki, who watched a nest in Panama for most of eleven days. Unfortunately, I cannot adapt the results of her vigils to table 2 because they were presented in graphical rather than tabular form. The male of Oniki's pair took sessions of up to 236 minutes; the female, up to 217 minutes. Together, they incubated for 88 percent of the daylight hours, which was almost the same as the performance of my pair of Chestnut-backed Antbirds in Costa Rica.

Although when abroad in the forest, especially when foraging with army ants, Spotted Antbirds are not particularly shy but rather more confiding in the presence of a human than most of their feathered associates, at their nests I have found them so wary that they steal inconspicuously away the moment a human comes into view—in this contrasting with all the foregoing antbirds, who sit steadfastly. Well concealed in a blind in a Panamanian forest, I watched a Spotted Antbirds' nest all one rainy afternoon and all the following forenoon. The diurnal sessions of incubation were long for such small birds; with a single changeover they filled the morning or the afternoon. The male sat for periods of 77, 44, and 217 minutes; the female, for 127, 36, and 164 minutes, plus her long nocturnal session of 11 hours and 35 minutes. Excluding this, the male incubated for a total of 338 minutes, the female for 327 minutes, and the eggs were unattended for only 54 minutes, or 7.5 percent of the daytime. These birds approached their nest by flying from slender sapling to slender sapling, to whose vertical stems they clung about a foot above the ground. Coming to incubate, the male called *chip chip chip chip*. As they snuggled down on the eggs, both parents repeated these notes, then continued to sit in silence.

As a male White-flanked Antwren flew down to the rim of his low nest, he displayed his usually concealed narrow white shoulder bands. He could sit continuously for three hours, and so bravely that I could

Black-and-White Antbird *Myrmochanes hemileucurus* Male Eastern Ecuador to northern Bolivia and Amazonian Brazil

almost touch him. His mate, who incubated for periods exceeding two hours, would not permit me to approach so near. Despite their ability to endure long fasts, these small birds incubated far less constantly than any other antbird that I have watched at the nest. A record covering all of one afternoon and all the following morning revealed that they attended their nest for only 76 percent of twelve diurnal hours. Such inconstancy was so unexpected that I watched this nest again for part of the following morning, and another nest for an entire morning, with much the same results. At the second nest, the eggs were incubated for only 50 percent of six hours.

On each of the three mornings that I watched at these nests, the bird who had passed the night on the eggs (doubtless the female) flew off while it was still too dark to distinguish its sex, and thereafter the nest was left unattended for an hour and a half to nearly two hours, until the female (twice) or the male (once) came to cover the thoroughly chilled eggs. Some of the intervals of neglect later in the day were almost as long. Apparently, the cause of White-flanked Antwrens' desultory incubation is their strong habit of foraging with the mixed-species flocks that pass through their territory, as none of the other antbirds whose incubation we have considered does more than occasionally. Several times the incubating White-flanked Antwrens abandoned their eggs when they heard a distant antwren's voice, and flew

TABLE 3. *Incubation and Nestling Periods of Antbirds*

Species	Incubation period (days)	Nestling period (days)	Locality
Great Antshrike	17–18	12–13	Costa Rica
Black-crested Antshrike	14	—	Suriname
Collared Antshrike	15	11	Ecuador
Black-hooded Antshrike	14–15	10–11	Costa Rica
Barred Antshrike	14	12–13	Costa Rica, Suriname
Slaty Antshrike	14	9	Panama
Plain Antvireo	15	9 (10)	Costa Rica
Streaked-crowned Antvireo	14 (15?)	11	Costa Rica
White-flanked Antwren	16	8	Panama
Slaty Antwren	15	—	Costa Rica
Dotted-winged Antwren	—	9	Panama
Dusky Antbird	—	11	Costa Rica
Chestnut-backed Antbird	—	10	Costa Rica
Bicolored Antbird	15–16	13–15	Costa Rica, Panama
Spotted Antbird	15–16	11–12	Costa Rica, Panama
Black-faced Antthrush	20	18 (17)	Costa Rica
Variegated Antpitta	17	—	French Guiana
Rusty-breasted Antpitta	16–17	—	Venezuela
Rufous Gnateater	—	14	Brazil

Note: Less usual periods are given in parentheses.

off as though to join it. In spite of irregular incubation, the eggs in the first nest hatched in sixteen days; those in the second nest were lost.

Our survey reveals that male antbirds' sessions on the eggs tend to be longer than those of their partners, and that in the hours between the termination of the female's nocturnal session in the morning and her return to her nest for the night, the male's total time on the nest often much exceeds hers. However, the disparity of the parts the sexes take in daytime incubation is not always as great as appears in table 2, for often the female begins her nocturnal session many minutes before other diurnal birds go to roost, or prolongs it after they have become active at daybreak. But it is difficult to determine with precision the length of a certain bird's active day.

As I passed through a Panamanian forest at daybreak, I noticed a nest that I had not previously seen. The open cup was filled with a loose mass of brownish feathers, some of which spilled over the rim. From one side of this mass some tail feathers stuck out, but in the dim light of an overcast dawn I could distinguish neither head nor bill. It appeared that a bird had died and disintegrated in the nest. However, when I touched the structure, the heap of loose feathers contracted and darted away so swiftly that, in the obscurity, I could not identify the bird who bore them. Returning later in the day, I found a female Slaty Antshrike sitting in this nest with her contour feathers laid flat in their usual diurnal position. On another night, I again found this antshrike sleeping in her nest with feathers so loosely spread that they seemed not to be attached to a living body; and in subsequent years I have seen Dusky and Bicolored antbirds sleeping on eggs with their plumage similarly relaxed—a habit that may be widespread among antbirds. I know no bird of any other family that sleeps on its nest with its plumage in such disarray. Whether when roosting amid foliage antbirds spread their feathers so wildly, I do not know. Allen Young's detailed study of a roosting Spotted Antbird makes no mention of this striking phenomenon.

We know the incubation periods of only a few of the many species of antbirds (table 3). Most range from fourteen to sixteen days. This is two to four days longer than the periods of songbirds (oscines) of similar size, despite the fact that, by sharing incubation, both sexes of antbirds keep their eggs more constantly covered than is usual among songbirds, in which incubation by males is exceptional. The few records for antpittas indicate that their eggs take a day or two longer to hatch, perhaps because their nests are so flimsy, as do the eggs of Great Antshrikes, apparently because of their larger size. The longest known incubation period of any antbird, twenty days, is that of the Black-faced Antthrush, whose eggs in a deep cavity appear to be safer than those in the open nests that most antbirds build.

7 *The Young & Their Care*

Early in the morning of June 2, 1937, while the first egg of a pair of Black-hooded Antshrikes was hatching beneath the male, his mate arrived quietly and perched about a yard away. With the feathers of her crown raised into a low crest, she called with subdued notes. Although the male had already been sitting for an hour and a half, he would not relinquish the nest to her, as he usually did. After a minute she flew away to chase some trespassing bird that I did not see well. The incubating male had become restless, often rising up to look down into the nest. At 8:56, a few minutes after his mate's departure, he picked up the cap of an eggshell and carried it away. I took advantage of his absence to slip from my blind and raise a mirror over the nest, which was above my head. While I viewed the reflected image of a newly hatched nestling, its father returned so promptly that he caught me in the act and protested loudly. After I hurried back to concealment, he returned to remove the large part of the empty shell. Within minutes he was at the nest again, stood on the rim to look intently down into it, then brooded.

At 9:24 the female called softly from nearby and was answered by her partner in low tones. When she called again, five minutes later, he left the nest. Continuing to voice soft notes, she approached the nest by flitting from twig to twig. Reaching the rim, she delayed there for several minutes, silently contemplating her first nestling, which she now saw for the first time. After she had been sitting for only seven minutes, the male, at 9:42, returned with a particle of food in his bill and perched near her. The female neither offered to take it from him nor made way for him to deliver it directly to the nestling. Continuing to hold the morsel in his bill, he lingered nearby for five minutes, then advanced to a point close behind the female, who left. Promptly he gave the nestling its first meal, fifty-one minutes after he had removed the first piece of the shell from which it had just emerged. After feeding, he brooded.

At 11:10, the female antshrike called nearby. Answering with low notes, her partner jumped from the nest and flew toward her. Soon she came into view, calling rather loudly, and at 11:15 gave the nestling its second meal, delayed two minutes looking down at it, then brooded. Two points in this account are of special interest. The first is the male parent's reluctance to relinquish the nest to his consort while the first egg hatched beneath him. The second, as I have learned while watching closely at the nests of many passerines, is his neglect to inform her that she had become the mother of a nestling, a fact that she did not learn until she saw it with her own eyes. This failure of communication is widespread even among birds that cooperate closely in nest attendance. The sight of her offspring prompted the female to bring an insect for it on her next return to the nest. The second egg hatched in the afternoon of the same day.

Like most passerines, newly hatched antbirds are helpless and blind. Most of those that I have seen bore no trace of feathers on their pink, dark flesh-color or, as in the Dusky and Spotted antbirds, black skins. Exceptional are hatchling Black-faced Antthrushes, covered with dark gray down unusually dense for a passerine bird. Newborn Variegated Antpittas in French Guiana, described by C. Erard, had black skins with tufts of dusky brown down on the feather tracts. Scaled Antpittas a few days old, found by Alden Miller in Colombia, had dark skins rather heavily covered with black down. Hatched with naked, orange-flesh-color skin, the Rusty-breasted Antpittas studied by Paul Schwartz in Venezuela were clothed in fluffy down when ten days old. In this they differed from most other antbirds, whose first plumage consists of flat, expanded contour feathers. The interior of hatchling antbirds' mouths is usually yellow or orange-yellow, but in the Spectacled Antpitta it is bright orange, with paler orange flanges at the corners. The empty shells are promptly carried away by the parents.

Nestling Black-hooded Antshrikes, Chestnut-backed Antbirds, and Black-faced Antthrushes were brooded by day in decreasing amounts until they were six or seven days old; but on the morning when a

Great Antshrike was nine days old, it was covered three times by its father for a total of 64 minutes, and twice by its mother (after her first early-morning departure from the nest) for a total of 40 minutes, during 5 hours of watching. This big antbird develops more slowly and remains in the nest longer than the smaller species. As they incubated the eggs through the night, so females brood their nestlings. By day males tend to brood more than females, just as they incubated more. In 66 hours of watching at four nests of Dotted-winged Antwrens, Greenberg and Gradwohl timed 16 sessions by males for a total of 470 minutes, and 25 sessions by females for 293 minutes. In 26 hours at the Black-hooded Antshrikes' nest where I watched hatching, the male brooded for 702 minutes and the female for 484 minutes. However, on the morning when the two nestlings were seven days old, their mother covered them for 135 minutes, their father for only 8 minutes. In 17¼ hours at a nest of Chestnut-backed Antbirds, the male brooded for 435 minutes and the female for 210 minutes. At this nest only the father brooded on the morning when the nestling was six days old, taking four turns that totaled 98 minutes.

Both parents feed the nestlings, bringing them grasshoppers, crickets, cockroaches, beetles, other adult insects, caterpillars, spiders, sowbugs, and the like. Large antbirds sometimes give their nestlings small lizards. A favored item is the fat abdomen of an insect, from which head and thorax, with all their appendages, have been removed by the parent before it approaches the nest. Very seldom have I seen them bring a berry. Antbirds carry their nestlings' food conspicuously in their bills, usually one item at a time, at least when it is large. To minimize activity at a nest, they bring large but infrequent meals. The average hourly rate over an extended interval rarely exceeds three meals per nestling, and often it is much less (table 4). In a prolonged study of a Rufous-throated Antbirds' nest in Brazil, Oniki learned that nestlings two to six days old were fed at the average rate of 0.7 times each per hour. Then the rate increased in the mornings to 1.7 to 2.2 feedings per hour until the nestlings were eight or nine days old, after which it dropped to 0.5 to 0.9 meals per hour for each of them until they were thirteen or fourteen days of age. At many nests, male antbirds bring food more frequently than do females. Many small insectivorous birds provision their young at much faster rates than antbirds do.

TABLE 4. *Rate of Feeding Nestling Antbirds*

Species[1]	Nestlings			Number of feedings			
				By			Per nestling-hour
	Hours watched	Number	Age (days)	Male	Female	Both	
Great Antshrike	5	1	4	3	2	5	1
	5	1	9	7	6	13	2.6
Black-hooded Antshrike	6.5	2	1	4	3	7	0.5
	6.5	2	4	5	5	10	0.8
	6.5	2	7	19	12	31	2.4
Slaty Antshrike	7.3	2	5, 6	11	7	18	1.2
Plain Antvireo	3	2	9	8	8	16	2.7
Streaked-crowned Antvireo	5	2	6	4	3	7	0.7
White-flanked Antwren	4	2	4	7	6	13	1.6
Dotted-winged Antwren[2]	9.75	2	0–5	9	13	22	1.1
Dusky Antbird	12.5	2	4–6	19	37	56	2.2
Chestnut-backed Antbird	11.5	1	1–3	19	6	25	2.2
	11.75	1	6–8	31	3	38[3]	3.2
Bicolored Antbird	6.25	2	1	7	4	11	0.9
Bicolored Antbird[4]	53.6	2	1–13	79	62	141	1.3
Black-faced Antthrush	4	1	10			7	1.8
Black-faced Antthrush	4.4	2	17			5	0.6

[1] When the name of the antbird is not repeated, the records are all for the same nest.
[2] From Johnson 1953, Panama.
[3] The sex of the food-bringer was not always recognized.
[4] From Willis 1967, Panama. All the other records were made by the author in Costa Rica, except that for the White-flanked Antwren, which was made in Panama.

When necessary, antbirds can greatly accelerate their rate of bringing food. A pair of Dusky Antbirds, mainly the male, remained away from their nest for surprisingly long intervals, detained, I surmised, by some excitement off in the forest. For one period of 101 minutes, the two nestlings were wholly neglected. Becoming hungry and restless, they shook their swinging cradle and called in weak voices that sounded far away. When at last their mother arrived, she fed them and cleaned the nest. In the next 24 minutes, she brought five meals. Soon their father appeared with an insect and began to help appease the nestlings' acute hunger. In 69 minutes they were fed 17 times by the female and 6 times by the male, a total of 23 meals, or at the rate of 20 times per hour for the two of them, which was more than four times the average rate of 2.2 meals per nestling-hour that I recorded at this nest. When feeding most actively, the two parents brought food 9 times in a quarter hour.

In the interval of accelerated feedings, the insects brought to the Dusky Antbirds' nest averaged much smaller than when the parents' visits were more widely spaced. Apparently, they now brought the first edible items that they found, whereas at other times they promptly ate the smaller ones, reserving the larger ones for their young. At the end of the busy interval, the nestlings appeared to be satiated, and their parents remained absent for the next half hour. When the young do not promptly swallow their food, the parent passes it from one to another until it disappears, or the parent finally eats the unaccepted offering.

Both parents remove the nestlings' white fecal sacs, often swallowing them at first, carrying them away in their bills after their progeny are older. Probably the adults' more efficient digestion can extract nourishment from the droppings of nestlings only a few days old. The manner of disposing of fecal sacs by Black-hooded Antshrikes was not related to brooding. At first, after feeding a nestling, the parent swallowed a dropping and remained to brood. A few days later, it sometimes carried away a sac in its bill, then promptly returned to brood.

As we have seen, most antbirds hatch with naked skin and tightly closed eyes. They develop rapidly. A Spotted Antbird less than a day old partly opened its eyes when it stretched up its gaping mouth for food. The eyes of other species begin to open when they are two or three days old. A day after hatching, the buds of pinfeathers are visible through the transparent skin of head, body, and wings. A day later, the pins are erupting from the skin, those of the remiges most advanced. These feather sheaths elongate rapidly until eight-day-old nestlings bristle with them. Then the feathers shed their horny sheaths, beginning at the tips, and so rapidly that at the age of nine or ten days the nestling is fairly well clothed in expanded plumage. This, at least, is the course of feathering in antbirds that never develop a garment of fluffy down.

I watched the departure of two young Black-hooded Antshrikes from their nest. When I entered the blind on a drizzly dawn, the nestlings were alone. If their mother had brooded them through the night, she had slipped from the nest while it was still too dark to see her. At 5:47, as daylight increased, the father came, singing in an undertone. Hearing his voice, the nestlings raised their gaping mouths before he came into view. He fed them and departed. The ten-day-old antshrikes preened their newly expanded plumage. Three times more their father fed them, and twice he carried away droppings, before their mother appeared with a large green insect.

The nestling who became feathered first (probably the one who was older by half a day) was restless, standing up and moving around in the nest. A few minutes later it climbed out to perch on the supporting limb close beside the nest, called in a small voice, and preened vigorously. It resembled its father in plumage, but its tail was stubby. After three minutes, it hopped back into the nest. In the next twenty minutes, it was fed twice by its father and once by its mother. At 6:51 the older nestling again hopped from the nest to rest beside it. Arriving once more with food, the father alighted by the nest, shaking the supporting twig and upsetting the nestling, who fell almost straight down into the tangle of low bushes and vines below. Its father instantly followed it downward; his response was so rapid that the two appeared to fall together, the parent slightly above the fledgling, as we must now call the young bird. Both were immediately lost to view amid the foliage. After a few minutes the parent reappeared, but I did not again see the fledgling. He took another insect into the bushes where his progeny was hiding before both passed beyond my ken into the dense vegetation at the forest's edge.

Through the rest of the day, the mother alone attended the young antshrike who remained at home, calling weakly *tit tit tit tit* when hungry, and making a low, buzzing sound when it stretched up to take food. She fed it eight times between 7:15 A.M. and 12:05 P.M., and six times between 2:40 and 5:40 P.M. Once, while removing a fecal sac, she carelessly dropped it from her bill, dived down in pursuit, and spent several minutes searching for it in the tangle. In the morning she did not brood, but after delivering a meal she occasionally lingered beside the nest, guarding. During the hardest of the afternoon's intermittent showers, she brooded, and continued to cover the nestling while the forest dripped after the downpour ceased, for a total of forty-five minutes. But she left the nestling uncovered through the night that followed, although now in June nocturnal rains were frequent.

The following morning I watched the second nestling leave. Before sunrise its mother, who had slept at a distance, approached through the underwood, calling loudly *wek wek wek wek*. The nestling answered with a soft *tow tow tow*, rapidly repeated, and two minutes later received its breakfast. In the next half hour, it was fed twice more. At 6:20 it hopped from the nest to an arm of the supporting fork, where it called, preened, and stretched its wings sideward and upward. After a few minutes the fledgling sidled along the twig to the main branch. Soon it lost its balance, fell, caught hold of a leaf, and clung precariously until it lost its grip and tumbled down into the bushes and vines. Its mother, who was resting nearby, promptly followed it to the ground, much as the father had followed the first young bird. She led the fledgling silently away through the undergrowth.

Two young Spotted Antbirds watched by Willis departed their low nest in much the same way as the Black-hooded Antshrikes had done. In midmorning, when both parents called nearby, one nestling answered with a loud *peeee*. When an adult moved to a low palm beneath the nest, a nestling hopped up on the cup's rim, then fluttered to the ground, where its father led it away. The mother continued to feed the remaining nestling, and brooded it through the night that followed. Next morning she slipped off the nest in the dim light of dawn, and fed the nestling eleven times in a half hour. After the last of these meals, she stayed below the nest. After several trials, the nestling succeeded in hopping up to the rim, from which it promptly leaped to the ground below her. At two other nests of Spotted Antbirds, only one parent attended the nestling who remained after the departure of its sibling. At one of these nests, the stay-at-home was cared for by the male parent; at the other, by the female.

The first young Black-hooded Antshrike severed contact with its nest 9 days and 22 hours after its father removed the eggshell from which it had just escaped. Its sibling left when between 10½ and 10¾ days old. Spotted Antbirds depart the nest at the age of 11 or 12 days. More precocious are White-flanked Antwrens, one of whom departed when only 8 days old, and Slaty Antshrikes and Plain Antvireos, who leave at nine days. Other antbirds for which we have information stay longer in the nest, up to 12 or 13 days in the Great Antshrike, and usually 18 days in the hole-nesting Black-faced Antthrush (table 3).

In Black-hooded Antshrikes, Spotted Antbirds, Bicolored Antbirds, and probably many other species, broods of two are divided between the parents, each taking full charge of one fledgling, apparently the fledgling who left the nest when that parent was present. Brood division is widespread among birds, especially those that nest near or on the ground, and promotes more efficient parental care. Amid dense ground cover, it is easier for a parent to keep in contact with a single young bird than with several; and separation of the brood probably reduces losses to predators. One advantage of having a brood of only two is that each of the fledglings can receive a parent's full attention. However, this will not explain why many tropical birds lay only two eggs, for broods of two manakins, hummingbirds, some flycatchers, and others have only a single parent, the female, to attend them.

Willis learned that if one young Spotted Antbird is lost, the attendant parent only infrequently helps its mate feed the surviving sibling. However, after a hawk killed a male, his widow took charge of both their offspring. In six families of which the female attended a single surviving fledgling, her mate sometimes gave her food that she passed to it. If, as happened eighteen times, the male was in charge of the only offspring, his consort begged from him but received nothing until the young bird started to feed itself. Apparently, male Spotted Antbirds feed single young about thrice as often as females do because they take greater care of nestlings during their last few days in the nest, and are more likely to be present when the nestling departs. Occasionally, mates permanently exchange their dependents; in one case, the father took his daughter; the mother, her son. However, in broods with one young of each sex, Willis noticed no significant tendency for the parents to take charge of the fledgling of the opposite sex. In eight such broods, males fed young males, and in eleven families females fed their sons.

Rufous-rumped Antwren *Terenura callinota* Male Panama and northern South America

While a fledgling Spotted Antbird waits quietly, its attendant parent wanders about seeking food. When it captures a suitable item, the parent chews it and starts to sing. The young bird answers with faint peeping, looks around, and flicks its tail. If the fledgling does not answer, the parent sings more and more loudly and searches until it finds its offspring, who flutters and squeaks as it stretches its open mouth toward its attendant. The parent pushes the insect into the mouth, watches until it is gulped down, then flies away. To lead its fledgling onward, a parent advances a few yards ahead, chirping and singing, twitching its tail, and frequently turning to look back. The young bird follows, hopping and fluttering over the ground or among low twigs. When they have reached a secluded spot, the parent feeds the fledgling. Sometimes an adult brings meals to its young from an ant swarm a hundred or even two hundred yards away, but more often it forages near its dependent.

Spotted Antbirds attain full adult size and tail length during the third week after they leave the nest, or when they are about a month old. At this age they are in full postjuvenal molt, and by the end of their second month they are difficult to distinguish from adults of their sex. When five weeks old, some young Spotteds begin to forage clumsily for themselves, but their discrimination is poor and they pick up such inedible objects as fragments of twigs or leaves, even whole dead leaves, and they peck at small protuberances on branches. By the eighth week after hatching, the young forage for themselves even when they accompany their parents. However, the adults, becoming increasingly interested in each other, now reject their offspring, snarling at and supplanting them, especially those of the parent's sex. All ten-week-old Spotted Antbirds have separated from their parents.

Juvenile antwrens remain longer with their parents. Young Checker-throated Antwrens disperse between one and three months after they leave the nest. White-flanked and Dotted-winged antwrens live throughout the year in groups of two to six individuals, all in excess of two consisting of parents with offspring who have stayed with them. A male Dotted-winged Antwren begged noisily for four months. In family groups of Dotted-wings, males are more numerous than females. In the forest of southern Costa Rica, a flock of one female and four males, all in adult plumage, kept close company from late October until at least mid-December. Greenberg and Gradwohl found immatures of these three species of antwrens closely associating with adults who were starting to nest again or had already raised a later brood. Such prolonged attachment of parents and offspring was most frequent in White-flanked Antwrens, nesting pairs of which were sometimes accompanied by both fledglings and older young, but rare in Checker-throated and Dotted-winged antwrens. Young of the latter usually disappeared before their parents laid in a new nest. Despite their frequent presence at active nests, older offspring of none of these antwrens have been known to serve as helpers.

As earlier told, female Bicolored Antbirds rejected by their parents are subordinate at ant swarms and have difficulty procuring enough food. In this predicament, they beg from bachelor males, who frequently feed them, thereby beginning an association that may ripen into wedlock. Both parent Black-spotted Bare-eyes attend their single fledgling who, when a male, may stay with them for several years, feeding his mother and engaging in nestsite-showing displays, carrying bits of nest material. Occasionally, he feeds young of another family. Although the young male Bare-eye is held aloof by his father, when his family is threatened by another family it draws more closely together, perching wing to wing.

The most closely knit families known among antbirds are those of the large Ocellated Antbirds, who in lowland forests from eastern Honduras to northwestern Ecuador live in enduring clans, composed of the parent pair, their sons and their sons' mates from other families. Up to six of these strikingly attired birds with bare, azure-blue faces keep close company, assuring one another of their continuing presence by constantly repeating contented little trills, sometimes whispered, rarely loud, while they flag their long black tails up and down. At ant swarms, where they cling to thin upright saplings, their size and unity enable them to hold the most productive central locations. Possibly, Ocellated Antbirds and Black-spotted Bare-eyes breed cooperatively, but despite much searching, the nests of these two species have eluded naturalists.

After they leave their low nests when about fourteen days old, Rufous Gnateaters probably stay with their parents until they are about two months of age. Then, for up to eighty days more, they forage alone in a corner of their parents' territory.

8 *The Black-faced Antthrush*

I have studied antbirds mainly in a tract of mature rain forest on our nature reserve around 2,400 feet (730 m) above sea level in the valley of El General, on the Pacific slope of southern Costa Rica. In this forest a bewildering diversity of trees form a canopy about 125 feet (38 m) high, above which tower giant palos de vaca, or cow trees (*Brosimum utile*), so called because their copious white latex was, in pioneer days, used as a substitute for cow's milk. Another lofty tree is the jacaranda (*Jacaranda copaia*), which in March covers its convex crown with a glorious array of lavender blossoms, but like many swiftly growing trees, does not live long. In April and May the mayo (*Vochysia aurea*) spreads golden bloom over its glossy foliage. Lianas spiral up trunks to display their efflorescence in sunshine with that of the trees. With slender gray trunks supported on wide-spreading, spiny prop roots, chonta palms (*Socratea durissima*) soar upward a hundred feet (30 m) to hold rosettes of great fronds in the canopy. A vast variety of ferns, aroids, bromeliads, orchids, and small epiphytic trees burden older trunks and more massive boughs. In deep shade beneath the trees grow the saplings that will eventually replace them, shrubs especially of the melastome and madder families, small spiny palms, tall tree ferns and lower ferns, large-leaved heliconias, and terrestrial aroids.

In this tract of about a hundred acres (40 hectares) of old forest, openings made by fallen trees in its midst, forest edges, and adjoining second-growth woods and thickets, I have over the years found fourteen species of antbirds. One, the Immaculate Antbird, is a rare visitor from higher altitudes. Two, the Great Antshrike and Barred Antshrike, dwellers in thickets rather than forests, formerly nested here but disappeared as surrounding lands were more intensively planted

Immaculate Antbird *Myrmeciza immaculata* Male Northern Honduras to western Ecuador and eastern Colombia

with coffee, sugarcane, and other crops. I have found and studied nests of all these antbirds except the Immaculate and the Bare-crowned, which apparently have never been described. I did not expect to find the Immaculate breeding at this altitude, but the Bare-crowned is with us throughout the year. The black male with naked, bright blue crown and face, which his rufous mate lacks, is so different in appearance from other antbirds that I greatly desired to learn more about their habits, but much searching has failed to disclose a nest.

Walking through this forest, I hear the triple whistle of the Black-faced Antthrush more often than I glimpse the bird in the deep shadows of the undergrowth. Occasionally, however, it emerges into the path ahead, giving a unique opportunity to watch an unusual bird, a passerine that resembles a small rail. Deliberately, the long-legged bird walks over the leaf-strewn ground, tilting forward with each step, its short black tail held erect. About seven inches (18 cm) long, it has deep brown or olive-brown upperparts, becoming chestnut on the upper tail coverts. Its face and throat are black, with a small white spot in front of each eye. Below, the stout-bodied bird is dark gray mixed with olive, with light brown or tawny under tail coverts. The back and sides of the neck are chestnut or cinnamon-rufous. Behind each large, dark eye is a crescent of bluish white bare skin which, in an erect head, makes the bird appear alert. Its short, straight bill is black; its long legs and toes pinkish with a dusky tinge. The Black-faced Antthrush is so different in many ways from all the other antbirds that I know, or that have been carefully studied by other naturalists, that it deserves a special chapter.

Often, seeking food, the antthrush moves in wide circles over the forest floor within its territory. Unless hard-pressed, it prefers not to use its short wings but walks quietly off into the shadows. Nevertheless, the little "cock-of-the-woods," as this bird is called in Trinidad, can fly strongly for considerable distances, as it often does when leaving its nest in a hollow stub. I have only twice seen an antthrush perch in a tree. As I walked along a woodland trail, one of these birds rose from the ground to a slender, horizontal branch above my head. Here it rested for about fifteen minutes, seeming not to notice me standing a few yards away. Sometimes it twitched its short tail up and down, or wagged it from side to side, but mostly it stood immobile. After a while it flew to a higher branch, delayed there for a few minutes, then rose to a third perch about fifteen feet (4.6 m) above the steep, wooded slope. Finally, it dropped to the ground and walked away. Several years later, an antthrush perched about five yards (4.6 m) up on a thin horizontal branch while it continued to answer whistled imitations of its song. Between repetitions it preened. After a while it flew down on a long, descending course.

Antthrushes forage alone; if they retain mates throughout the year, they do not keep close company with them. I told in chapter 2 how they walk sedately over the ground, flicking small leaves away with a sideward movement of the bill, picking up larger ones and tossing them, eating the insects and other small invertebrates they thereby disclose. Although not "professional" army-ant followers, like many other birds they take advantage of the easy foraging at the swarms they happen to encounter on their daily rounds. Instead of perching above the horde, as most ant followers do, these terrestrial birds avoid ant stings by walking around the outskirts of the throng, capturing fugitives that seem to have the best chance of escaping. Occasionally two, rarely three, antthrushes attend an ant swarm, without staying close together. Sometimes, like dominant professionals, they drive smaller birds away from productive spots. Their visits to ant swarms

Black-faced Antthrush *Formicarius analis* Sexes similar Southeastern Mexico to northern Bolivia and eastern Brazil

are usually brief, but occasionally one remains for most of a day. From Costa Rica to Amazonia, Willis found Black-faced Antthrushes at raids of *Eciton burchelli,* less often with the smaller swarms of *Labidus praedator.*

Rarely, as I roamed through the forest, an antthrush has flown with an explosive *tleet* from a hollow stub close beside me, thereby revealing the site of its nest. Over the years, largely by accident, I have found seven such sites,—three in palms, three in small, dead dicotyledonous trees, one in the hollow trunk of a living dicotyledonous tree. The highest nest was in a stilt palm stub with a jagged top. From the pinnacle, 14 feet (4 m) above the ground, the opening extended obliquely downward to 11 feet (3.4 m), the lowest point through which the birds could enter and leave. The eggs lay 13 inches (33 cm) below this lowest part of the aperture, in a well-like hollow about 4 inches (10 cm) in diameter.

Another nest in a stilt palm was in a strange situation. A section of a tall trunk had fallen away and stood on the slope below, held nearly upright by a single loop of a dead frond of a twining fern, which permitted it to sway precariously when I touched it. Reduced by the decay of its internal tissues to a thin and slightly flexible shell, this segment of palm trunk was a tube nearly 7 feet long and 5½ inches in internal diameter (2 m by 14 cm). In the uphill side of the tube, 5 feet (1.5 m) above its base, was an irregular opening with jagged edges, which served as the antthrushes' doorway; nearer the ground were several holes too narrow for the birds to pass through them. To approach or leave the nest at the bottom of the hollow cylinder, they had to climb up or down almost 5 feet.

Even lower was a nest in a hollow stump, apparently of a wild papaya (*Carica* sp.) that I found in Costa Rica's Caribbean lowlands. The thin-walled stump, about 3 inches (7.5 cm) thick at the top, was only 18 inches (46 cm) high, but the hollow extended through the taproot so far into the ground that the eggs rested 14 inches (35.5 cm) below the surface. If this subterranean nest had not been situated on a well-drained slope, it would probably have been flooded by June's heavy rains. A few days after the birds began to incubate, some passerby pulled up the flimsy stump, leaving the eggs exposed in the bottom of the resulting hole. As I approached, an antthrush who had been sitting in the altered nest rose up through the ground litter and flew away. Next day the eggs were abandoned.

The first nest that I found was in the slightly leaning trunk of a dicotyledonous tree, far advanced in decay, about 12 feet high and 7 inches thick (3.5 m and 18 cm). It was covered with green moss, and here and there a few aroids and ferns grew upon it. In the side of the trunk, 6 feet (1.8 m) above the ground, a large gap gave access to the central hollow. Beside this a smaller gap extended slightly lower and was the preferred doorway of the antthrushes. The eggs rested about 2 feet (60 cm) below the opening, or about 4 feet (1.2 m) above the ground. This was the only nest site that was not open at the top, and it offered shelter from rain. The site that received most sets of eggs and most study was a dicotyledonous stub about 9 feet high and 5 inches thick (2.7 m and 12.5 cm). Both the open top and a long, narrow slit in the side were used as doorways by the occupants. An aroid and a species of *Carludovica* crept over this slender trunk which, despite its decrepitude, stood for three seasons to contain seven nesting attempts, five of which were successful.

Whenever there was a prospect of finding a later brood in a cavity, I did not disturb it. I examined the nests in only three, one of which had been torn open by a mammal. The bottom of each cavity was filled with dead leaves, leaf skeletons, and strips of palm fronds, all of which served to raise the eggs nearer the entrance. Upward, this accumulation of leaves was interspersed with slender rachises of compound leaves, thin petioles, flower stalks, or long, slender flowers too decayed for identification. At the top, these materials, unmixed with leaves, formed a thick bed on which eggs and nestlings rested. Even in the same cavity, the level of this bed fell as the supporting mass of leaves decayed or, when the cavity was occupied repeatedly, rose as more material was added for a new brood. No nest was less than 13 inches (33 cm) below the opening used by the antthrushes, and one, as already told, was 5 feet (1.5 m) down. The drop from entrance to eggs of twelve nests averaged 30.8 inches (78.2 cm). The birds could pass through an aperture only 1½ inches (3.8 cm) wide. An unpleasant odor of decay emanated from these nests resting upon sodden, decaying leaves, for the antthrushes nested in months when rain fell almost daily.

Between March and September, I found eleven sets, each of two eggs. Since I could not reach these eggs without opening the cavity, I viewed them in a mirror inserted through the entrance, while the chamber was illuminated by an electric bulb attached by a wire to a flashlight. So consistently did the eggs appear to be white, finely speckled and flecked over the whole surface with dark brown, that I long believed this to be their original color. Careful washing of an abandoned set convinced me that antthrushes' eggs are plain, unmarked white. Soon after they are laid, they become heavily soiled, apparently with earth from the birds' feet. The absence of shell markings raises a problem. Plain white eggs are laid by birds, such as woodpeckers and kingfishers, that consistently nest in holes or burrows—the primary hole-nesters. Hole-nesting species of families in which they are exceptional, such as the American flycatchers, often have heavily pigmented eggs, apparently an inheritance from ancestors that built open nests. Since the majority of antbirds have open nests, why do antthrushes lay plain white eggs? I surmise that natural selection eliminated pigment from their eggs because this, plus the dirt that soon covers them, would make the eggs so dark that the parents could hardly see them in dimly lighted hollows, with the result that they too frequently broke them.

As I saw on repeated vigils, the antthrush who incubates through the night, probably the female, leaves her eggs while the dawn light is still dim in the forest. Often she stands in the entrance and utters a long series of whistles before she flies away, or she calls sharply *tleet* as she takes wing. The nest remains unattended until one member of the pair, probably the male, arrives, on the mornings that I watched, from 12 to 44 minutes later. Thereafter, I found the eggs constantly incubated through the forenoon by the parents sitting alternately for intervals of rarely less than 2 hours and sometimes as long as 5 hours. The oncoming partner arrives by walking over the ground until near the base of the stub that contains the nest, a mode of approach that might reveal lurking enemies before showing the nest's location by entering it. Often the newcomer flies up and in before the incubating partner leaves. Since the sexes look alike, in these cases the watcher cannot be certain that the mates have exchanged places, but this is a fair assumption. At other times, the bird who has been incubating flies out when it hears its partner approach. The changeover is made without sounds audible a few yards away, but occasionally the outgoing bird sings loudly in the doorway before it departs. I have not watched nests for long hours during the mostly rainy afternoons of the nesting season, but I have sometimes found the eggs unattended early in the afternoon, as they were early in the morning.

Antthrushes show much discrimination in exposing themselves in their doorways. Most of my nests were revealed to me by a startling rush of wings, accompanied by sharp *tleet tleet tleet* calls, as I walked unsuspectingly by the stubs that contained them; but with other animals antthrushes are more cautious. While a squirrel scolded nearby, then jumped upon the frail stub within which an antthrush was hidden, the bird remained discreetly out of sight, thereby avoiding the betrayal of its eggs to a nest-robbing rodent. On another day, the antthrush did not leave its nest when an agouti bounded by with three howling dogs in close pursuit, and I shouted at the trespassing hounds to drive them away. Yet this same bird peered through the doorway when an Orange-billed Sparrow hopped over the ground near the stub, and its mate looked out when a small lizard crawled by. Slight, rustling sounds, such as might be made by the approach of snakes, their chief enemies, alerted them. Sometimes tapping on the stub fails to make them reveal their presence within. On several occasions, an antthrush stuck to its eggs while I lowered a lighted bulb above its head, then rushed out past the light and my eyes.

At one nest, both eggs hatched twenty days after the last was laid. At another nest, both hatched eighteen days after I found the full set. No other antbird that I know has such a long incubation period. The empty shells are promptly removed by the parents.

The nestling Black-faced Antthrushes surprised me. Hatchlings of a dozen species that I had already seen were all quite naked; but as soon as antthrushes dried, they became a featureless mound of dark gray down, exceptionally dense for a passerine bird. Except for the possibly greater diameter of the mound, two nestlings presented almost the same appearance as one, so closely huddled were they. To learn their number was not easy. But if I continued to look down at them in a lighted chamber, after a while I would see one move slightly. Its sibling, when two were present, would make a small compensating movement to bring itself again into closest contact with the other. These swift motions often revealed flashes of pink on a nestling's neck, and perhaps a momentary exposure of its abdomen. Occasionally a bill or two projected slightly from the fluffy mound.

As the nestlings grew older, the chief difference that I noticed was that the dark mass spread farther over the bottom of the well. Rarely were pinfeathers visible amid the down before the nestlings were fifteen days old, when I sometimes glimpsed the sheaths of the wing plumes. The nestlings' most conspicuous features were the wide, flaring, whitish flanges at the corners of their mouths. The projection at the base of the upper mandible did not fit tightly against that at the base of the lower mandible, but between them was a gap that resembled the horizontal pupil of an eye. As I peered down into the dimly lighted tube, each nestling appeared to be staring up at me with great white eyes. The effect was sometimes startling. I surmised that, in addition to guiding food-bringing parents to receptive mouths, the eyelike flanges might frighten predators and so be doubly useful. No other nestling antbird that I have seen has such prominent oral rims. The antthrushes' true eyes were much less evident than these false eyes in front of them. Once I glimpsed partly open eyes on a nestling three days old, but I rarely noticed the eyes of a nestling antthrush of any age.

Not only were the young antthrushes nearly featureless, they were nearly always silent and inactive when I visited them. At all but one of the nests, it was nearly always impossible to elicit any vocal or other response from them. The exception was the nest in the pipelike fallen section of palm trunk that swayed when it was touched. When I looked in at these nestlings three days after they hatched, they stretched their heads far upward with widely gaping mouths that displayed the yellow interior between the whitish corners. They made a buzzing sound, which after a few days changed to a sort of chiming-sizzling chorus. It appeared that these nestlings stretched up for food when I shook their palm trunk because they could not distinguish this movement from that imparted to it by the arrival of a parent. But at the other, less mobile stubs, the nestlings evidently learned to respond to stimuli that I could not reproduce, such as the whirr of approaching wings or the scratching sounds made by the adults' toes as they climbed down the shaft. A nestling two weeks old emitted several low, froglike grunts when my mirror knocked against the side of the doorway, but I could not later elicit these notes by the same means. Aside from these instances, I rarely heard the nestlings on my many visits to them.

As blackish nestlings lying inert at the bottom of a dark cavity seemed likely to escape notice by predators that hunt by sight, or possibly to frighten by their staring false eyes one that looked in on them; so by their habitual silence, contrasting with the loquacity of certain other nestlings, they avoided attracting enemies that hunt by sound. When a parent arrived with food, the young became noisy; but the adult approached the nest only after careful scrutiny had failed to disclose lurking peril. A ten-day-old nestling greeted each parental visit with a continuous whirring sound, which began when it heard the rustle of approaching wings and continued until the parent flew from the doorway. Only when nearly ready to fly did nestlings receive their meals with loud, liquid *chips*, which they also voiced when they heard an adult's notes in the distance.

My highest nest was in the midst of tangled vegetation that filled a gap in the forest made by the fall of the great tree that broke off the top of the chonta palm and converted it into a nest site. Here, where the ground was cluttered, the parents bringing food arrived on the wing, from a point beyond my vision, and shot into the hollow trunk with never a pause at the opening. After fifteen or twenty seconds, they would appear at the orifice, delay there a few seconds, then fly outward and downward until lost to view.

A nest more satisfactory for watching the antthrushes attend their young was situated in a dell beneath magnificent tall trees. Here, where the ground was fairly free of obstructions, the parents usually approached by walking. Sometimes one of them continued for about five minutes to march slowly around the stub, as though carefully inspecting the surroundings. One day a parent, arriving with food, circulated around and around the blind in which I sat, frequently within a yard of it, eyeing it intently, at first repeating at short intervals a loud, full double note, but afterward in silence. For nearly twenty minutes, he continued to scrutinize my blind from all sides. Aside from a Bicolored Antbird, no other bird has exhibited such obvious curiosity in the triangular enclosure of brown cloth that concealed me.

The terrestrial approach of these antthrushes provided excellent opportunities to see what they brought for their nestlings. Unfortunately, the food was nearly always badly mangled before the birds came into view. The most recognizable object was a single small lizard. Usually the food appeared to be insects, but their mutilated state prevented recognition of their kinds or the number brought on a single visit. Sometimes a wing or two dangled from the amorphous mass in the parent's bill. Once this mass was whitish, with dark wings hanging loosely from it. Similar masses without revealing wings were more frequent. I surmised that they came from the viscera of insects. Even with laden bill, the adult often continued to push aside the ground litter, searching for more food as it marched toward the nest. Usually it came and left silently.

Meals were infrequent. In the first four hours of a morning, a single ten-day-old nestling was fed seven times. In the first four and a half hours of another morning, two seventeen-day-old nestlings received only five meals, each of which appeared to be very substantial. After the nestlings were a week old, I did not find the parents brooding them by day. Since I never saw them remove a dropping in the bill, yet they disappeared from the cavity, they must have swallowed the fecal sacs. From the age of three days, if not earlier, the nestlings attached these prominent white bodies to the wall of the cavity beside themselves. As the nestlings grew older, they deposited them higher on the wall, well above the bottom where they rested. The adults continued to remove this waste until, a day or two before the departure of the young, they climbed up to take their food through the doorway. No longer finding it necessary to enter the cavity to deliver meals, the parents relaxed their attention to sanitation, and droppings accumulated on the wall just inside the entrance—a sign that the young were about to depart.

The young in my first nest were taken by a predator when a few days old. From the third nest, in a hollow palm stub, the single nestling vanished when about eighteen days old. I found no indication of predation, but on the day before it disappeared it looked as downy, as lacking in flight feathers to bear it away, as when it hatched. Did it have, hidden beneath its long, loose down, expanded wing feathers to ease its descent to the ground? Or did it simply drop from the top of the stub and walk away, in the manner of downy Wood Ducks and other precocial birds that nest at a height? These questions remained unanswered until I studied my fourth nest, five years later.

I entered the blind before this nest at dawn of the first day after I found the single nestling's droppings stuck to the wall at the entrance. Soon I heard the calls of an antthrush approaching through the still dusky forest. Looking around for the parent, I almost missed the event that I most desired to see; at that moment the fledgling emerged from the hollow stub. It flew well on a long, descending course and alighted on the ground a good distance from the nest, where it called loudly *chip chip chip*. It looked much like the adults. When I tried to catch it for closer examination, it retreated unsteadily over the rough ground by a combination of walking and hopping, and when I came almost within reach it flew too well to be captured. During the chase, a parent followed us, calling sharply *tleet tleet tleet.* Finally, the fledgling walked into a tangle of bushes, vines, and fallen boughs where I could not follow it.

The departure of the first brood in the following year was somewhat different. When the two nestlings were eighteen days old, I entered the blind at daybreak to watch. Soon I noticed, standing silently on the ground in front of me, a parent who had approached so quietly that I did not notice its arrival. The light was still too dim to reveal whether it held food in its bill. Two minutes later, it flew up to the top of the stub. Then I saw three shadowy figures fly down in swift succession, and I heard a rapid *chip chip chip,* followed by loud, full notes of the parent. These notes grew fainter as the family walked off through the dimly lighted undergrowth, where I could not see them. Going to the nest with light and mirror to confirm my observation that both fledglings had flown from it with their parent, I found it empty.

A nestling of the third brood in 1959 vanished when only seventeen days old, apparently having left spontaneously. Around sunrise next morning, a parent silently fed the remaining young at the doorway, then flew down. After hesitating a few seconds, the fledgling followed, calling as the others had done. Alighting on the ground, it moved around as though bewildered, then started to walk away from its loudly calling parent. In less than a minute, it recovered from its confusion and came toward the parent, who was now behind me. Passing right through the blind, within two inches of my foot, the youngster joined the adult, and the two walked away together.

The six fledgling Black-faced Antthrushes, of five broods, that I watched leave their nests departed when eighteen days old, or rarely a day younger or older. Every one of them flew from its nest cavity before sunrise, when a parent arrived with the day's first meal. If the parent called as it approached the nest, the young bird flew out to meet it. If the parent arrived silently, the fledgling, probably after receiving food at the doorway, flew down with the departing parent, or sometimes it followed after hesitating for a few seconds. The young antthrush flew from the nest, often rather strongly, but after alighting on the ground it walked away with the adult. The departure was always accompanied by continuing loud *chip*s from the fledgling.

Although the parent's early morning arrival was the signal for the fledgling's departure, the former apparently did nothing to induce the latter to leave; severance from the nest was determined primarily by the young bird's development. Its emergence in the dim light of early morning appeared to be related to the incubating parent's habitual departure from the nest at this hour. Like the Black-hooded Antshrikes that I watched leave their nests and the Spotted Antbirds that Willis watched, the antthrushes always departed with a parent who led them away. Possibly this is usual in antbirds; we have barely begun to learn the habits of this fascinating tropical family.

After a chase, I caught a young antthrush who had just left its nest and appeared somewhat retarded, perhaps as a consequence of the heavy rains that fell into its roofless cavity while it was growing up in August. The eighteen-day-old fledgling retained much down on its upperparts, especially on the back of its neck and shoulders. Dark gray in color, the downy plumes were about an inch (2.5 cm) long and richly branched. Beneath them, expanded contour feathers completely clothed the body. Although slightly duller than adults, the young antthrush closely resembled them. Its dusky remiges were fairly well expanded, but its tail was very short. Its bill was black with a white tip, and the white flanges at the corners of its mouth were still prominent. The eyes were dark brown. Its strong legs and toes were dark flesh-color; the toenails grayish horn-color. After examining the fledgling and making notes, I replaced it in its hollow stub, where it remained until it heard a parent calling in the distance in the following dawn. The nineteen-day-old fledgling flew more weakly from the nest and walked away more slowly than the others when a day younger.

9 *Parental Devotion*

Antbirds are strongly attached to their nests and young and appear greatly distressed when their progeny are jeopardized. At the first nest of any antbird that I ever saw, that of a pair of Slaty Antshrikes on Barro Colorado Island, sixty years ago, the parents repeatedly demonstrated their admirable devotion. This nest was eight feet (2.4 m) up in a horizontal fork of a small tree, beneath a forest giant with wide-spreading plank buttresses. After I had watched the birds build and incubate by day, I wished to learn which parent sat through the night. When I arrived at five o'clock in the evening, the male was on the nest, and with his usual staunchness remained until I gently lowered the supporting branch to make certain of his identity. A few minutes after he flew away, the female came silently to settle on the eggs. Here she sat until the organ notes of a distant Great Tinamou heralded the approach of night.

After dusk had settled deeply in the underwood around me, I approached the nest for a final close inspection, for with a novice's caution I wished to be quite sure that the female was still present. As I turned to leave, a stick beneath my foot broke with a sharp report that frightened her from the nest. Repenting my blunder, which I feared would cause the eggs to remain unwarmed through the night, if they were not permanently abandoned, I withdrew a few paces and watched. In a few minutes, the brave little bird came hopping back to the nest that she could hardly see. After another interval, I switched on my flashlight and saw her white-tipped tail sticking up above the nest's mossy rim. Birds of other families have permanently deserted their nests with less provocation. This, my first experience of an antbird's

Slaty Antshrike *Thamnophilus punctatus* Male incubating Guatemala to Bolivia and southeastern Brazil

extraordinary attachment to its nest, also taught me to make inspections of this sort not at the night's beginning but at its end, when the risk of causing desertion is less.

The Slaty Antshrikes' strong fidelity to their nest was demonstrated in a different way when I photographed them on it. While I attached my old-fashioned ground-glass-plate camera to a ladder, and with my head beneath a black cloth focused it on the nest, the male continued to flit around me, coming within two yards, and complaining almost constantly. His more timid partner remained slightly more aloof. When the camera was ready, I retired into a blind of palm fronds that I had built between the buttresses of the neighboring tree—an unnecessary precaution. Hardly had I entered this nook when the male settled on the eggs, a yard in front of the camera's glassy eye, and I pulled the thread that released the shutter. After each change of film, one parent or the other promptly returned to the nest, with the result that in little over an hour I took five clear photographs, three of the male and two of his partner.

A few days after the nestlings hatched, I found their mother resting on the nest's rim, looking into it, inspecting or guarding them. When I was little more than an arm's length away, she dropped to the ground and beat her wings against the carpet of dead leaves as she fluttered slowly and apparently painfully away in a convincing "crippled bird" act. I followed until she had lured me onward about twenty feet, when she rose into the nearest bush and called. Almost immediately, her mate appeared with a nasal *churr* and advanced closer to me than she had dared. When I returned to the nest to examine the nestlings, he stood on a twig beside it in a belligerent attitude, wings spread and fluttering, tail expanding into a dark fan bordered by white spots, black crown feathers erect and bristling, and the feathers of his back turned outward to reveal a broad central patch of snowy white, the presence of which I had not suspected. When I started to lift a nestling from its pensile cradle, the guardian father lunged forward and bit my finger. Doubtless all the strength of outraged parenthood went into that nip, so weak that I hardly felt it. Two more swift attacks on my fingers were followed by equally swift retreats, before the antshrike withdrew a short distance. Meanwhile, the less courageous mother perched in a neighboring bush and complained with a nervous rattle, all the while displaying a hitherto concealed patch of white in the middle of her olive-brown back.

At only one other of the nests of hundreds of species of birds of diverse families that I have visited have the parents simultaneously used different means to divert me. When I placed a hand palm downward over the nest of a pair of Gray Catbirds in a barberry hedge in a suburban garden in Maryland, the female bit my fingers with her slightly deformed bill, while her watchful mate buffeted the back of my head. This double attack was often repeated, but the female would never peck my hand when I laid it on her nest palm upward; she appeared to know how it worked.

Another antbird whose parental zeal overruled prudence was a Bicolored. As I passed along a narrow trail through our forest, one of these antbirds flew up and clung to a slender vertical stem close in front of me. Holding in its bill a large, fat insect, it repeated a low, scolding *churr* over and over. When I turned around to look for the nest that I suspected was nearby, the bird darted past me, so close that it brushed my leg with a wing. Again it clung in front of me and complained. While I searched through the bushes beside the trail, it dropped to the ground and beat its half-spread wings against fallen leaves, repeating this act in various spots a yard or two from me. Soon I discovered two nestlings in pinfeathers, about five or six days old, lying helpless on the ground, as though they had been tumbled there. Both were cold and sluggish; one had fresh blood on a flank, from a slight abrasion of the skin. They appeared to have come from a nearby palm stump, so fragile that it had collapsed, possibly under the weight of the attendant parents.

The behavior of the single parent, of undetermined sex, then present was amazing. It gulped down the insect that it had been holding. With the clearance of its mouth its utterance changed to an incessantly repeated *p'r-r-r-r-r* that sounded like an angry protest, while it continued its distraction displays. Each was performed in a single spot, where with stationary body the bird struck its partly spread wings against the ground, and lasted a few seconds. I placed the cold nestlings in the palm of my hand and held it on the ground in front of the groveling parent, who advanced and bit a finger, not once but three or four times.

Spreading my handkerchief over the fallen leaves, I laid the nestlings on it, while I pondered what to do with them. The adult sat or lay on the ground, facing the handkerchief from less than a yard away, watching or guarding the nestlings for the ten minutes or more that I left them there. Since I had read that Bicolored Antbirds sometimes nest in the sheathing base of a large palm frond, I cut the sheath from a great fallen leaf of a stilt palm and set this stiff, hollow cylinder, open along one side, upright against the spiny prop roots of the palm tree, close by the collapsed palm stump. On a stuffing of leaves in the bottom of the yardlong cylinder, I placed the nest that had fallen from the stump, and covered the open top with a green leaf. Then, while the parent, still resting on the ground two or three feet away, intently watched me, I laid the nestlings, one by one, on the nest. I could do no more for them.

Would the antbirds attend and keep alive their progeny in this improvised nest? During the hour that I stayed to watch, neither brought food nor visited them, although I heard the voices of Bicolored Antbirds in the distance. Next morning, both nestlings were dead.

Along the same woodland path in the following year, in a low, hollow palm stump similar to the one that had collapsed, I found my first intact Bicolored Antbirds' nest and passed many hours studying it. As already told, the parent incubating or brooding in this nest would return my gaze while I bent over it. If I made it jump out by waving my hand above it or gently tapping on the frail shell, it would grovel in a distraction display almost at my feet, then rise and circle around me a few yards away, clinging near the ground to slender upright stems, and repeating over and over a slightly churred *p'r-r-r-r p'r-r-r-r.* On the day I found this nest, the antbird hovered close around me, calling and watching me intently, the whole time I was engaged in examining the nest and making notes. Rarely have I seen a bird evince such obvious concern over what was happening to its nest and eggs. Finally, becoming bolder, it advanced within arm's length while I bent over the stump, then shuffled away a few inches with mincing steps, trying to entice me to follow.

When I finished my notes and started to walk away, the antbird followed me for several paces. Then it returned and peered down into the stump, as though to see whether its eggs were still there. After I had proceeded several yards, I stopped to watch. Within five minutes, the bird entered the hollow to resume incubation. When, during my study, I found that I could not distinguish the parents without marking them, I remembered that Jimmy would permit me to ruffle his plumage with a stick. Cutting a slender wand, I fastened a tuft of cotton on its end, soaked this in paint, and readily touched the antbirds with it while they hovered, complaining, around me, leaving on their plumage spots of color just large enough for identification.

I have not seen an antbird interact with any mammal except myself and a vegetarian agouti. Willis watched Bicolored Antbirds flutter their wings on low saplings or the ground in front of an omnivorous coati who approached their nest. One of the parents alighted on the quadruped's back and pecked it, causing the animal to snap back blindly at its assailant. Despite this resolute defense, the coati might have plundered their nest if it had not been deterred by big, stinging ants that emerged when the animal sniffed with its long snout at the base of the nest cavity.

After sitting in their nests until I came within reach, both male and female Chestnut-backed Antbirds have given earnest distraction displays, fluttering over the ground with spread, beating wings, or at times holding them nearly erect as they crept away. One of their nests was situated sixteen inches (40 cm) up in a twining fern that had overgrown a small, spiny palm, beneath a towering palo de vaca tree, in a part of the forest with little ground cover. The tree was freely shedding large seeds that attracted agoutis.

Early on the first morning that I watched, while the solitary nestling was less than one day old, one of these rodents passed close by the nest without appearing to notice it. Less than a yard away, it sat on its fat haunches, lifted a seed in its forepaws, and ate. The male antbird came near the mammal, churred, hopped over the ground nearby, flicked a leaf aside with his bill, and brooded. While he sat in the nest, the agouti, continually twitching its pink ears to keep off swarming mosquitoes, rested, then ate more seeds nearby. Presently, it walked all around the nest, scarcely a foot from it, and sniffed at the structure with its blunt nose. At this point, the male antbird jumped from the nest and flew low over the ground, while the agouti, possibly startled by the bird's sudden movement, bounded away in the same direction. From the undergrowth beyond my view, I heard the antbirds' scolding *wittit wittit wittit.* Soon the rodent returned to its former position near the nest but gave no more attention to it, and after a few minutes the animal wandered away.

For the next week, agoutis continued to eat palo de vaca seeds around the nest, while the antbirds fed and brooded their nestling, birds and rodents ignoring one another, as far as I saw during two all-morning vigils. But late on the third morning, when the eight-day-old nestling's feathers were beginning to escape from their long sheaths, two agoutis walked toward the nest while the male antbird was hopping away on the opposite side. The parent promptly returned, dropped to the ground in front of an agouti, who had come close to the nest, and fluttered away as though injured, causing the animal to start in his direction, whereupon the bird flew up beyond reach of the earthbound rodent. After following for a foot or so, the agouti stopped and wrinkled its nose to smell the air near the nest. The antbird continued incessantly to voice the alarm note, *wittit wittit wittit wittit,* punctuated at intervals by the call *waaa.* Again and again he dropped to the ground and simulated injury in front of the mammal, without succeeding in luring it away from the nest.

Although on earlier days, when the antbird mostly ignored the agoutis, none of them did more than occasionally smell the air around the nest, his demonstrations seemed to excite a stronger interest in it.

Now one of them sniffed and sniffed, wrinkling its broad nose while moving all around the little clump of plants that sheltered the nest. It advanced closer and closer until it touched the structure with the tip of its nose—whereupon I violently shook the cloth that surrounded me, making the animal bolt away with the usual sneezelike bark of alarm. It left the nest slightly tilted. Soon it returned and went toward the nest with a directness that left no doubt of its destination. Bursting from the blind, I chased it off through the forest.

After the agouti departed and I reentered the blind, the male antbird returned, hopped over the ground beneath the nest, and clung low on upright stems until he found a small insect that he took to the nestling. The whole time the agouti was present, the female remained out of sight. At break of the following day, I found her brooding the lone nestling in the nest, which I had propped up after the agouti tilted it. I was obliged to be absent most of the day. When I returned in mid-afternoon, I found the nest torn from its supports. The nine-day-old nestling had vanished; whether it had escaped or been taken by an agouti or some other animal, I could not tell.

The antbird's distraction displays, which might have lured away a predator, such as a dog, a tayra, or opossum, with the prospect of a meal, failed to entice the agouti from the vicinity of the nest. Although almost wholly vegetarian, these rodents lick the glair from inside freshly emptied shells of hens' eggs that we throw out to them from the kitchen, which suggests that they would not be averse to a small bird's egg. However, their long neglect of the antbirds' nest around which they ate *Brosimum* seeds showed that they are not frequent nest robbers—although there might be individual differences in this respect. The final destruction of the nest by an agouti, if one of them was in fact responsible for it, appears to have resulted from the male Chestnut-backed Antbird's excessive zeal, which drew the rodents' attention to a nest that they might otherwise have ignored.

Realistic "broken-wing" displays are also given by Plain Antvireos. After clinging to a head-high nest until an intruding human has come almost within reach, the bird drops straight to the ground before it begins its act, often appearing to spring from my feet. The first time I saw a Plain Antvireo's distraction display, I searched over the ground before I learned that I had been standing with my head close beside its nest. One female varied her performance by clinging, an inch or two above the ground, to the sapling that held her nest and slowly fluttering her spread wings. Then she crept slowly and apparently painfully over the ground, pausing frequently to beat her wings against it.

Instead of following a straight course away from the nest, which would have taken him beyond view in dense undergrowth, a male antvireo continued to double back and forth close by me, more effectively holding my attention. Once this same male prolonged his act for about a minute. Plain Antvireos make their displays more conspicuous by exposing usually concealed bands on their shoulders, white on the male, buffy on the female. If it fails to lead the intruder away by its demonstration, the antvireo rises to a branch above its head and protests with low, liquid, mournful notes—*cher, cher, cher, cher*—repeated as long as the unwelcome visitor remains. Sometimes the other parent arrives and joins in this melodious dirge.

A pair of Dusky Antbirds whose nest hung above my head, beside a rivulet at the forest's edge, remained steadfast in their deep pouch until I shook the tangle of climbing razor sedge that supported it. Then the male would drop down to a log and stand there briefly, exposing a gleaming patch of white feathers in the middle of his dark back, without simulating injury. A female White-flanked Antwren would cling to her low nest, brooding her nestlings, until I came very near. Then she would jump from it and drag herself over the ground for many yards, convincingly conveying the impression that she was severely injured and frantically trying to escape. After she had led me a good distance from her nest, she flew up into the bushes and joined her mate in uttering the queer calls characteristic of these antwrens.

Distraction displays tend to be unpredictable. One member of a species, or of a pair, may give them when a human visits a nest, whereas another does not. Or a certain individual may display at one time but leave its nest inconspicuously at another time. Much appears to depend upon how long a parent has been sitting, and whether it has a clear stage for its act. Birds wisely refrain from such demonstrations amid vegetation that would impede their movements, increasing the risk of being caught. They act as though they are in full control of their limbs, not delirious or having a fit, as some observers have too hastily concluded. Indeed, few things that a bird does require such cool calculation, to avoid being caught if the enemy approaches too near, or to discourage its pursuit if they flutter too far ahead.

At nests of little Slaty Antwrens, individual variation in distraction displays was great. At some, both parents performed; at others, only the male or only the female. After sticking to her nest until I almost touched her, one female would drop to the ground and act like a crippled butterfly. After sitting firm above his nestlings until I came within reach, a male would drop almost straight down to cling, a few inches above the ground, to a thin stem, where he slowly waved widely spread wings, revealing at the junction of each with his body a small white patch, ordinarily invisible, that contrasted with his dark plumage. When I approached him, he would flutter from sapling to sapling ahead of me, repeating his striking act, trying earnestly to lure me on. Sometimes he waited until I was only a few feet from him before he retreated to another perch. The farther he led me from his nestlings, the briefer his performances became. When I was far enough away, he vanished in the undergrowth and called his mate.

At one Slaty Antwrens' nest, the parents' behavior changed as incubation advanced. I first came into view of this nest in the forest when I emerged from dense undergrowth into a fairly clear space about sixty feet (18 m) from it. When the eggs were newly laid, the antwrens slipped from their nest the moment they caught sight of me, so unobtrusively that I rarely glimpsed them. About the middle of the incubation period, their behavior changed abruptly. Now they would watch me advance in full view until I came close to them, then drop from the nest and display. They had shifted from one to another of two prudent alternatives. In the first, they abandoned their nest while the intruder was distant and unlikely to notice their departure; in the second, they stuck to it until they were sure that they had been detected, then tried to lure him away by feigning injury. To have left the nest when a potential enemy was in full view, but possibly had not yet noticed it, would have gratuitously called his attention to it by the movement of departure. Could these small heads have learned this, or was their changed strategy innately determined?

Of the antbirds that I have studied, the Spotted Antbird and the Barred Antshrike have least often continued to sit while I approached their nests. Frequently it was difficult to glimpse them on their eggs or nestlings. As Willis found, Spotted Antbirds are most likely to give distraction displays, exposing the white patch on their backs, when their fledged young appear to be menaced. Parent Barred Antshrikes stayed prudently out of sight whenever I visited their nestlings. The cries of a feathered nestling that I held in my hand drew its mother from the surrounding thicket. With her long rufous crest erect, and uttering at intervals a liquid, mournful call, she circled around us through the bushes, at a distance of five to ten feet (1.5 to 3 m). Her cries, added to those of her nestling, drew her partner, who came with his loose, black-and-white crest standing straight up, and likewise circled around us, voicing a harsh, long-drawn, nasal complaint. Neither adult tried to drive or lure me away, as the smaller Slaty Antshrikes so valiantly did. Perhaps Barred Antshrikes, who inhabit thickets, light woods, and gardens in cultivated districts, are more wary because they have had longer racial experience of how dangerous humans can be than have Slaty Antshrikes in wild forests.

Natural selection tends to set limits to the parental devotion of birds and other creatures. Not the parent who loses its life defending offspring who will perish without its care, but the parent who remains alive and can try to replace the brood it has lost, contributes more young to its species and sets the prevailing temper. Nevertheless, parental zeal sometimes overcomes instinctive caution, as frequently occurs in the antbirds most strongly attached to their progeny, at least when their nests are visited by students of bird behavior or photographers who might, if so inclined, catch or kill them when they come so near. The plaintive cries that Plain Antvireos, Barred Antshrikes, and other birds emit when their progeny are jeopardized, like the harsh, grating notes that many utter in angry or aggressive moods, suggest by their tone quality the emotions that would stir us in similar situations.

10 Nest Success, Breeding Seasons, & Survival

Although antbirds' nests are difficult to find and carefully guarded, most are destroyed by the many predators that lurk in tropical forests and thickets, mainly snakes. One rarely finds and learns the outcome of enough nests of a single species for a statistically significant statement of their success. Much of our information on this subject was gathered by Willis during his prolonged studies of ant-following birds. Of 84 Spotted Antbirds' nests that received eggs, only 16, or 19 percent, produced at least one fledgling. Of 76 nests of Bicolored Antbirds, 9, or 11.7 percent, yielded fledged young. Willis and Oniki believed that nest losses of 80 percent or more are frequent in the antbird family.

Such a high proportion of nest failures seemed excessive to me. Lacking a significant number of records of the outcome of nests of any one species of antbirds, I lumped all that I had made of whatever species. Over the years, I learned the results of 61 nests of 17 species of antbirds in which eggs were laid. Most of these nests were in the valley of El General, at 9 degrees north latitude in southern Costa Rica, a few in other parts of Costa Rica and in central Panama. Of the 61 nests, 25 produced at least one fledgling, giving a nest success of 41 percent. Most successful were those of the hole-nesting Black-faced Antthrush, 6 of whose 11 nesting attempts yielded fledged young, and of the Plain Antvireo, with 5 successes and 3 failures. Taken together, these 17 antbirds did somewhat better than other birds among which they lived.

Calculations of nest success are in most cases rough approximations; many factors influence their accuracy. If we include all nests in which eggs were laid, at whatever stage we found them, as I have done

Black-crested Antshrike *Sakesphorus canadensis* Male Northern South America

with these antbirds, our estimate may be too high, for nests with incubation far advanced, or with nestlings, are already a favored lot, having escaped perils that destroyed other nests in the vicinity. On the other hand, the estimate may be too low because well-hidden nests least likely to be found by the investigator may also be least likely to be pillaged by predators. Moreover, our visits may influence the outcome, as by leaving scent trails that keen-nosed mammals can follow, disturbing the foliage around nests, or causing protesting parents to behave in ways that draw attention to them.

The effect of human visits to nests has in recent years received considerable study. The conflicting conclusions reached by these investigations depended in large measure upon the species of bird, the locality, and the researcher's methods. Among the studies that revealed no adverse effects of human visits to nests was that of Willis, who by watching Bicolored Antbirds at army-ant swarms could tell when they had nests and whether they reared fledglings. This enabled him to compare the outcome of unvisited nests with that of nests that he periodically examined. Although he found no significant difference in the success of these two categories of nests, in both it was amazingly low even for birds in lowland rain forests. A few nests of Black-spotted Bare-eyes that Willis studied by the same indirect method were much more successful. Some of the studies made in the North Temperate Zone revealed that human visits do decrease the success of nests. Ordinarily, we cannot learn the outcome of nests without inspecting them at the critical periods of laying, hatching, and fledging; but the fewer our visits, and the more discreetly we make them, the greater the probability that these nests will produce offspring.

To compensate for their small broods of only two and their frequent failure to rear them, antbirds have long breeding seasons, during which they can try again and again to produce descendants. Within a few degrees of the equator, White-backed Fire-eyes near Belém, Brazil, and Black-crested and Barred antshrikes in northern Suriname nest, as species, throughout the year. Farther north, in Trinidad, twenty-two nests of Barred Antshrikes were well distributed through the year (table 5). In Panama and Costa Rica, at 9 degrees north latitude, from which most of our information on the breeding seasons of antbirds comes, most start to nest in March or April, when rains return after a short dry season, although they may anticipate the rains by a few weeks, and they continue to nest through much of the ensuing wet season, or over a period of six to eight months.

TABLE 5. *Breeding Seasons of Antbirds*

Species	Breeding season	Locality
Black-crested Antshrike	January–December	Suriname
Barred Antshrike	January–December	Suriname, Trinidad
Black-hooded Antshrike	February–August	Costa Rica
Slaty Antshrike	December–September	Panama
Plain Antvireo	February–July	Costa Rica
Plain Antvireo	April–August	Trinidad
White-flanked Antwren	March–August	Trinidad, Panama
Dotted-winged Antwren	February–September	Costa Rica
Dusky Antbird	February–October	Costa Rica, Panama
White-backed Fire-eye	January–December	Pará, Brazil
Chestnut-backed Antbird	April–October	Panama
Chestnut-backed Antbird	April–August	Costa Rica
Bicolored Antbird	April–January	Costa Rica
Bicolored Antbird	April–December	Panama
Ocellated Antbird	April–December	Panama
Spotted Antbird	April–November	Panama
Black-spotted Bare-eye	August–March	Pará, Brazil
Black-faced Antthrush	March–October	Costa Rica

Long-tailed Antbird *Drymophila caudata* Male Venezuela to Colombia and northern Bolivia

From April to December, some pairs of Bicolored Antbirds on Barro Colorado Island started six to eight nests, but none was known to bring more than one of them to a successful termination. Spotted Antbirds on the same island may nest three times between April and November; one pair raised two broods. In the valley of El General, where Black-faced Antthrushes nest from March to October, a pair raised three broods in this interval in the same tottering hollow stub, the best run of antbirds' luck that has come to my attention. Hole-nesting birds in general do better than those with open nests.

Tropical land birds that reproduce more slowly than their North-Temperate counterparts have longer life expectancy. In a study continued for six years on Barro Colorado Island, Greenberg and Gradwohl found an annual survival rate of 74 percent for White-flanked Antwrens, and of 62 percent for Checker-throated Antwrens. Willis and Oniki estimated an annual mortality of less than 20 percent for Spotted Antbirds, Rufous-throated Antbirds, and Black-spotted Bare-eyes, but more than 20 percent for Bicolored Antbirds, Ocellated Antbirds, White-plumed Antbirds, and White-backed Fire-eyes. Willis knew a Spotted Antbird who lived to be over ten years old. On Trinidad, David Snow and Alan Lill reported a banded male Plain Antvireo at least seven years old. A female antvireo, banded by Snow, was recaptured by Lill at the same place ten and a half years later. She was last found nesting when at least nine and a half years old, but none of her known nests was successful.

Male Bicolored Antbirds normally pair when over one year old. Females mate when only six months old and can lay eggs in the year they hatched, or when six and a half months of age. Little is known of the age of first breeding of other antbirds.

11 *Antbirds & Humans*

Living in wild forests, dense thickets, or bushy hedgerows in cultivated districts, antbirds have slight contact with humans. Taking little or no vegetable food, they neither attack field crops nor pilfer fruits from orchards. If they have any economic importance, this is probably in protecting timber trees from insects, a role that they share with so many other forest birds that it might be difficult to disentangle their contribution. I am aware of no study of antbirds' impact upon humans' economic interests. Fortunately for these birds, they are neither colorful nor songful enough to become popular as cage birds, and like other insectivores, they are difficult to maintain in captivity. Accordingly, they are largely exempt from the nefarious pet trade that has fallen so disastrously upon more brightly attired or melodious birds.

Humans' interest in antbirds is other than economic. They swell the lists of birders who visit tropical America to glimpse and jot down as many species as they can. Serious students of birds ponder their adaptations to life in tropical woodlands where raptors prey on adults and multitudes of snakes and small mammals plunder nests. Why do antbirds, like many other tropical birds from pigeons and hummingbirds to finches and tanagers, lay only two eggs, sometimes only one, whereas at higher latitudes clutches are usually larger? Why are their nestling periods so much shorter than their incubation periods, when in other small altricial birds these intervals tend to be more equal? How effective is the parents' defense of eggs and young, often so admirable when we visit their nests to study them, when directed against snakes and other nest-robbers? How long do antbirds live? Abundant in the lower levels of tropical forests, antbirds are favorable subjects for the study of these problems pertinent to tropical bird life in general.

Pale-faced Bare-eye *Skutchia borbae* Sexes alike Amazonian Brazil

Probably antbirds' major importance to humans, or at least to the small minority of them who wander through tropical forests for enjoyment or study, is aesthetic or intellectual. Magnificent tropical forests exalt the spirit while they humble us. Soaring trunks lift our gaze skyward, and often with it our minds. We marvel at the diversity of form of leaf and flower. The profusion of vegetation around us reminds us vividly of the exuberance of the creative energy. But in the deeply shaded underwood we find a dearth of the bright colors that delight us in sunny meadows and gardens. We are rarely long forgetful of lurking perils: venomous snakes and—in forests least altered by humans—jaguars and bands of white-lipped peccaries that might attack us. Falling branches or whole trees may maim or kill us. In great expanses of woodland, we may readily lose our way, and sometimes with it our lives. The narrowed outlook amid crowded trunks oppresses people accustomed to wider views; the magnitude of the trees makes us feel puny and ineffective; the prevailing silence may depress us.

A bird's song can dispel the loneliness that often overcomes us after long hours amid unbroken forest. We feel that some of the energy of sunlight absorbed by the luxuriant foliage above and around us should be enjoyed by creatures more sentient than plants appear to be. Forests seem to need animate life for their completion. Monkeys, agoutis, squirrels, and other mammals help to fill this need; if they do not menace us, we rejoice in their presence. But birds are the animals we are most likely to hear and see; more than the others, they make the forest feel less alien to us, a place where warm-blooded creatures that care for their families can thrive. Among the feathered denizens of the woodland, the numerous antbirds do much to make the forest seem friendly. The more intimately we know them, the more they attach us to their sylvan abode and make us feel at ease there.

Contemplation of the lives of antbirds raises searching questions. Do they feel anything like love or affection for the mates with whom they maintain yearlong or lifelong bonds? Are they emotionally attached to the nests and young that they protect so courageously? Does anger or fear surge up in them when their progeny are threatened? Do notes that sound so plaintive to us express genuine distress? Or do these active birds live in a psychic penumbra while they follow innate modes of behavior? Empathy often suggests positive answers to all but the last of these questions, but science remains stubbornly silent. To have trustworthy answers would be of the greatest philosophic importance. They would help us to interpret this enigmatic universe, and to assess the accomplishments of evolution. For only in the measure that it contains sentient beings who find joy or satisfaction in their existence can we ascribe value to the cosmos, and only if it enhances the psychic life of at least some of its multifarious creations can we find significance in the erratic course of evolution.

Part II
The Ovenbirds

12 *The Ovenbird Family*

In northern Honduras, long ago, I found Rufous-breasted Spinetails abundant in tangled thickets that swiftly covered neglected clearings amid the rain forest. I found many of their nests of interlaced sticks, and spent hours watching and photographing them, fascinated by structures so much bigger and more elaborate than the birds' nests that I knew in the northern land where I was born. The wren-sized builders of these avian castles went about their business unperturbed while I stood hardly more than arm's length away. I admired the care they took of their household, constantly hopping over it, inspecting its walls and crannies, tucking in more firmly twigs that were coming loose, pulling up pieces that were slipping down from the thatch that roofed the chamber where their eggs lay on a bed of downy leaves. After I made a small hole in the wall to examine a nest's contents, I closed it with little twigs such as the birds themselves used, but never to their satisfaction. Promptly they noticed the alteration I had made and proceeded to mend their wall more skillfully than I had done. Such attentive, capable birds!

Such was my endearing introduction to the ovenbird family (Furnariidae), with 214 species the fifth largest of the avian families confined to the New World. Of South American origin, ovenbirds spread over the whole continent, from its cold and stormy southern extremity to its warm Caribbean coast. From its southern homeland, the family extends in diminishing numbers of species through Central America to central Mexico, but not a single species reaches the United States or the Antilles. Two species have become established on the Juan Fernández Islands, five hundred miles (800 km) from the continent in the South Pacific, one on the Falklands, four hundred miles (650 km)

Thorn-tailed Rayadito *Aphrastura spinicauda* Sexes similar Southern Argentina and Chile

away in the South Atlantic, a number in Tierra del Fuego and the small islands around it, and five species in Trinidad, close by Venezuela. In South America, ovenbirds have adapted to almost every available habitat, from the coasts to the high, treeless páramos and puna, up to the edge of the snowfields on the highest Andean peaks. The record for altitude appears to be held by the Streaked-backed Canastero, which has been found up to 18,000 feet (5,500 m) on the volcanos of Ecuador, but several other species reach almost as high. From arid deserts to marshes and the rainiest forests, ovenbirds thrive.

In size ovenbirds range from about four to ten and a half inches (10 to 27 cm). The smallest is the Thorn-tailed Rayadito; the largest, the Brown Cacholote, both of southern South America. The sexes are always similar in plumage. In coloration ovenbirds are plain and monotonous, with shades of brown, ranging from rufous and chestnut to dusky, prevailing throughout the family. The underparts tend to be paler, buffy, gray, or whitish. Streaks and spots, variously distributed, help to separate species that are often confusingly similar. Blue, green, and pure red are absent from ovenbirds' feathers; small patches of orange (or near-orange) adorn the throats of several species, and yellow brightens those of the Yellow-throated and Chotoy spinetails. Exceptionally colorful is the head of the little Orange-fronted Plushcrown of western Amazonia. Among the most handsome ovenbirds is the eight-inch (20 cm) Pointed-tailed Palmcreeper, a little-known bird of northeastern South America. Its back, tail, and wings are bright rufous-chestnut; its head, nape, and underparts are black, regularly streaked or barred with white. Another admired species is the tiny Thorn-tailed Rayadito, with blackish upperparts, whitish underparts, broad buffy superciliary bands, and cinnamon wing bars. Its very unusual spine-tipped tail enhances its attractiveness.

Among the few crested ovenbirds are the Larklike Brushrunner, the Tufted and the Araucaria tit-spinetails, the three species of cacholotes, and the Crested Hornero. A tuft of long feathers adorns each side of the Buffy Tuftedcheek's head. In contrast to the rather uniform coloration of ovenbirds, their tails are so extraordinarily diverse that many are named for them: spinetails, thistletails, softtails, prickletails,

White-chinned Thistletail *Schizoeaca fuliginosa* Sexes similar Páramos, Venezuela to Peru

Des Mur's Wiretail *Sylviorthorhyncus desmursii* Sexes similar Southern Argentina and Chile

barbtails, wiretail. These tails range from short to conspicuously long, with rounded, pointed, or forked ends. The feathers that compose them have rounded, long-tapering, or abruptly pointed tips. As the birds forage through dense vegetation, their tail feathers become worn and frayed, making them appear more spinous. The resemblance of some of these tails to those of woodpeckers or woodcreepers suggests that they might serve as props when the birds that bear them climb up trunks. However, with few exceptions, their shafts are not stiff like those of woodpeckers, and their tips are not decurved, like those of woodcreepers. Moreover, many ovenbirds with sharp-pointed tails do not climb; and those that more or less frequently do appear to derive little support from pressing their soft tails against the trunk.

Strangely, the leaftossers, with tails so stiff and sharp-tipped that the birds have been named for them (*Sclerurus* = hardtail) are terrestrial and rarely cling to trees, apparently chiefly when they are alarmed. The most woodpeckerlike of the ovenbirds is the White-throated Treerunner of the beech (*Nothofagus*) forests of southern Chile and Argentina. It climbs up trunks and over branches, propping itself with its tail, each feather of which has a strong, stiff shaft that protrudes beyond the vanes in points which grip the bark. It so resembles the woodcreepers that it was formerly classified with them in the family (Dendrocolaptidae) most closely related to the ovenbirds.

Among the strangest tails is that of Des Mur's Wiretail, which lives in central and southern Chile and across the Andes in Argentina. The tail of this tiny bird consists of only six feathers, instead of the ten or twelve usual in ovenbirds. The almost vaneless central pair, about three times as long as the rest of the bird, are very slender and stiff. The next pair outward are thin like the middle pair but only about half as long. The outermost pair are rudimentary. The somewhat similar tails of the emu birds (*Stipiturus*) of Australia also have only six feathers, but they are much shorter and have loose barbs that impart a lacy aspect.

The rectrices of another very small bird, the Thorn-tailed Rayadito, have almost barbless ends, which project from the tail like so many sharp, stiff spines, up to half an inch (1.3 cm) long. Tit-spinetails of the genus *Leptasthenura* have long, strongly graduated tails, the central pair of which, projecting far beyond the others and tapering to sharp tips, sometimes make the tail appear forked. In the light of our present deficient knowledge of ovenbirds, the variation in their tails appears inexplicable. We do not know whether their strange forms are adaptations or vagaries of evolution.

Common Miner *Geositta cunicularia* Sexes similar Peru and southern Brazil to Tierra del Fuego

After attaining adult coloration, ovenbirds retain it throughout the year. Exceptional is the Straight-billed Earthcreeper, which in summer becomes paler above, less streaked below, than in its winter plumage.

Ovenbirds' bills are mostly short to moderately long, straight or slightly curved. Exceptionally long and downcurved are the bills of the Slender-billed Miner and the Buff-breasted Earthcreeper. The short bills of xenops have a straight upper mandible and a strongly upcurved lower mandible, which strengthens the bill for an upward thrust in a decaying twig. The bills of the Great Xenops and the two species of recurved bills are similar but larger. Ovenbirds have normal passerine feet, with three forwardly directed toes and one hind toe.

This very diverse family is divided into three subfamilies, of which the first is the Furnariinae. Among its members are ten species of miners of the genus *Geositta,* so called because they nest in tunnels that they dig into the earth. They are small, dull-colored, short-toed birds, mostly brownish and rufous, that run swiftly over the ground and fly reluctantly. Preferring sparsely vegetated, dry country, they are most abundant in southern South America and in the Andes from Peru southward, where they inhabit the grassy puna of Peru and stony slopes in Chile. At low elevations they live along the arid coast of Peru and Chile and interior deserts as well as the scrublands of Patagonia and Tierra del Fuego.

Earthcreepers of the genus *Upucerthia* resemble miners in appearance and habitat but tend to have longer bills and tails. They walk or run rather than creep over the ground, and fly swiftly but low. Like miners, they prefer open, rocky country with scattered bushes, but some species frequent lusher growth along streams and even wet meadows. Miners and earthcreepers appear to be the ecological counterparts of wheatears and stonechats of northern lands.

The eight species of cinclodes (genus *Cinclodes*) walk and run over the ground, but at least some of them perch in trees. They nest in burrows and crannies in open and rocky lands. Their geographical distribution is similar to that of miners and earthcreepers, but one species ranges northward to Colombia and northwestern Venezuela. Cinclodes prefer wetter habitats, the margins of rivers, ponds, and irrigation ditches, and stony slopes near streams. The Seaside Cinclodes inhabits rocky shores in Peru. As their name implies, cinclodes resemble dippers (*Cinclus*) in certain aspects of their lives, especially in their addiction to watercourses. They have also been compared to the wagtails

Scaly-throated Earthcreeper *Upucerthia dumetaria* Sexes similar Peru and Bolivia to Tierra del Fuego

and waterthrushes of the Northern Hemisphere. Closely allied to the foregoing genera is the Crag Chilia, the only species in the genus *Chilia.* Confined to the country for which it is named, it nests in craggy slopes of the central region.

At first sight, the six species of horneros of the genus *Furnarius* appear not to belong in the same subfamily with the burrow-nesters, for they build clay ovens above ground and some inhabit rain-forested regions, where they prefer river margins and forest edges. Although they walk over the ground, they are not so strictly terrestrial as miners and earthcreepers but often perch on trees, buildings, or wires. By making closed nests, they can thrive in regions where exposed banks and rocky outcrops are rare. A link with the miners and earthcreepers is the fact that all three nest in earth—but in earth that horneros have molded into a special form.

The largest subfamily, the Synallaxinae, is a very diverse assemblage of birds of varied habitats, from grasslands, marshes, and páramos to lowland and montane forests. Although many species forage on or near the ground, none is so terrestrial as are miners and their close relatives; they hop rather than walk. Some nest in holes and crannies, but most are builders of the wonderful nests that will claim much of our attention in later chapters. Long-tailed tit-spinetails have habits (other than nesting) not unlike those of the Long-tailed Tit (*Aegithalos*) of Eurasia, as suggested also by the specific name of the Plain-mantled Tit-Spinetail, *aegithaloides.* Hudson's Canastero has been compared to a pipit—again not in its nesting habits. Another resemblance to birds of different families is suggested by the name of the Larklike Brushrunner of southern South America, where larks are absent.

The third subfamily, Philydorinae, is also a diverse collection of genera, including the abundant foliage-gleaners, treehunters, barbtails, and tuftedcheeks, but also the terrestrial leaftossers. Here belong the curious little Plain Xenops, which has much in common with the tiny woodpeckers known as piculets, and also the White-throated Treerunner, with habits like those of a larger woodpecker, such as the Hairy or the Downy. This subfamily also contains the largely terrestrial, jaylike cacholotes, biggest of ovenbirds with the most spacious nests, and the Streamside Sharptail, which walks along the banks of South American watercourses. The large, handsome Pointed-tailed Palmcreeper adds distinction to this heterogeneous subfamily.

Most ovenbirds appear to be permanently resident wherever they occur and not much given to wandering, which probably accounts for the fact that, unlike those other great Neotropical families, the flycatchers and the hummingbirds, they have not colonized the temperate zone of North America. Certainly it cannot be cold weather that

Andean Tit-Spinetail *Leptasthenura andicola* Sexes similar Venezuela and Colombia to Bolivia

Spectacled Foliage-gleaner *Anabacerthia variegaticeps* Sexes similar
Southern Mexico to western Ecuador

has prevented their northward advance beyond the tropics, for they endure the harsh climates of Andean páramo and puna and, at low altitudes, Tierra del Fuego and the Falkland Islands, where the Blackish Cinclodes, there called the Tussac Bird, lives throughout the year.

Although no ovenbird (and few South American birds of other families) undertakes migrations comparable to those of hundreds of species that breed in the Northern Hemisphere, some of those that nest in the bleakest climates seek milder regions in winter. Among the inhabitants of the southern Andes that descend to lower altitudes are the Scaly-throated Earthcreeper and the Crag Chilia. Birds of high southern latitudes that migrate northward to winter in central Chile, northern Argentina, Uruguay, Paraguay, or southern Brazil include the Scaly-throated Earthcreeper, Bar-winged Cinclodes, Austral Canastero, Lesser Canastero, and Wrenlike Rushbird, which Hudson called "one of our few strictly migratory species in the family." Whether all these birds migrate by day or by night my books do not tell, but I surmise that they travel by easy stages in the daytime, as they do not need to cross great expanses of water or desert, like so many of the nocturnal migrants that breed in the Northern Hemisphere.

13 Food & Foraging

Ovenbirds are almost wholly insectivorous, using this term in the broad sense that includes not only winged insects and their larvae but also spiders and other small invertebrates. Some of the larger ovenbirds add a few small frogs and lizards to their diets, and a few species eat seeds, berries, and other vegetable foods. Ovenbirds' methods of foraging are as diverse as their bodily forms and their nests. In this chapter I treat them according to what appears to be the main mode of foraging of each genus or species, but many are too versatile to be limited to a single procedure.

Among ground foragers are the terrestrial miners, earthcreepers, Crag Chilias, cinclodes, and horneros. The first three are among the few members of the family reported to consume seeds and other vegetable products as well as many insects. On the high páramos of Colombia, the Stout-billed Cinclodes digs into soft soil, mud, and debris, flicking aside particles, as well as gleaning from the ground or foliage. On the Pacific coast from Lima southward, the Seaside Cinclodes stays close to the water's edge, where it forages on and around rocky promontories and the stony or sandy beaches of small coves between them. Advancing and retreating with the surf, it picks up small crustaceans, mollusks, and other marine animals, much in the manner of Purple Sandpipers. On the Falkland Islands the Blackish Cinclodes, or Tussac Bird, eats insects and their larvae, amphipods, and isopods left stranded with seaweeds on beaches when the tide recedes. Cracked eggs of Rockhopper Penguins diversify its fare, as do blowflies that parasitize sheep, which the cinclodes eats so freely that it keeps the small islands where it dwells almost free of them. A bold or tame bird, the Blackish Cinclodes enters houses to steal butter or cooked meat from dining tables, and it does not disdain carrion. It reveals its versatility by holding down food with its feet while it picks off pieces—a simple trick that many birds lack. In the same subfamily, horneros forage chiefly on open ground where they can walk unimpeded, picking up mature insects, larvae, and earthworms. The Pale-legged Hornero digs into soft soil with its bill.

Other ground-foragers are the Firewood-gatherer and the cacholotes. The Brown Cacholote rummages and digs among fallen leaves, bark, and other vegetable debris to expose small edible creatures, which it often holds with its big feet. It eats birds' eggs, including those of domestic hens, whom it terrorizes with loud screams and flashing wings until they flee from their nests. Seeds and buds torn from plants with its strong bill swell the fare of this largest of ovenbirds.

In humid tropical forests at low and middle altitudes, leaftossers (often miscalled leafscrapers) hop over the ground and squat on flexed legs, with their bills tirelessly tossing fallen leaves and other litter right and left, so vigorously that many pieces go flying a foot or more, and picking up the edible creatures that their strenuous, prolonged exercise exposes. A solitary leaftosser was so absorbed in its occupation that it continued for half an hour while I stood watching only a few

Stout-billed Cinclodes *Cinclodes excelsior* Sexes similar Colombia to Peru

Edge of rain forest beside a maize field in the valley of El General in southern Costa Rica, habitat of Scaly-throated Leaftosser, Buff-throated Automolus, Plain Xenops, Black-faced Antthrush, Bicolored Antbird, and Spectacled Antpitta

yards away. Occasionally, while I walked along a leaf-strewn forest path, a leaftosser has made the leaves fly ahead of me. As I advanced, it flew onward a short way and alighted to resume its leaf-tossing. In this manner we continued to travel together for about fifty paces, the bird moving when I moved, alighting to scatter leaves when I paused to watch. After a bout of leaf-tossing, the small brown bird rests crosswise on a low horizontal vine, branch, or fallen log. Along narrow, wooded watercourses in South America, the Streamside Sharptail (less appropriately called the Sharp-tailed Streamcreeper) walks deliberately beneath tangled vegetation, gathering food from the moist ground.

Among shrub-and-ground-foragers are many species of spinetails (*Synallaxis*), thornbirds (*Phacellodomus*), canasteros (*Asthenes*), and the Larklike Brushrunner. These birds hunt among bushes and thickets that are often so dense that it is difficult to watch the dull-colored foragers. They search branches, twigs, and foliage at no great height, peer into curled dead leaves, or rummage in accumulations of them. Often they descend to the ground, where they hop about investigating the litter, sometimes disappearing into piles of leaves or heaps of brush. At least one of these shrub-and-ground-foragers, the Rufous-breasted Spinetail, eats a few berries.

Moss-gleaners climb over mossy trunks and branches, searching for small creatures that lurk in the green vesture. In wet, mossy, epiphyte-encumbered forests at middle altitudes, and especially their darkest and dampest parts, the deep ravines into which only stray beams of sunshine find their way, the little Spotted Barbtail creeps over the trunks and branches of small trees and the lower regions of great ones, without ascending into sunny treetops. It appears to subsist wholly on insects and other invertebrates, in the quest of which it is so incessantly active that it is most difficult to keep in view as it works over mossy boles and branches in all attitudes, head or tail, back or belly, or either side uppermost. Although it sometimes ascends a vertical trunk with its body upright and tail pressed against it, in the manner of a woodpecker, on the whole it appears to depend little on its barbed tail feathers for support. Rarely it perches crosswise on a branch, in typical passerine fashion.

The Ruddy Treerunner is a moss-gleaner of similar habits that frequently forages like a rummager. The Red-faced Spinetail both moss-gleans and rummages, usually high in the trees of wet mountain forests. This little bird enriches its diet with the tiny white corpuscles

Pearled Treerunner *Margarornis squamiger* Sexes similar Northwestern Venezuela and Colombia to Bolivia

that accumulate on the hairy brown bases of the petioles of the great palmate leaves of cecropia trees, when the ants that subsist on them are absent—a dainty fare favored by many other small birds. Another ovenbird that includes vegetable material in its fare is the Itatiaia Thistletail, which eats many of the small drupes of a species of *Rapanea* of the myrsine family, as reported by Helmut Sick.

Rummagers forage mainly in dense tangles of vines and dead leaves lodged in them, or among large or massed epiphytes. One that I have often watched in rain forests at low altitudes is the Buff-throated Automolus (in bird guides inappropriately called "foliage-gleaner"). This large ovenbird is adept at clinging inverted, or in any other orientation that the situation demands, while with strong bill it assiduously probes folds of dead leaves or accumulations of them, in the underwood or up to midlevels of the great trees. The dead foliage of a recently fallen tree richly rewards its efforts. Holding a large item beneath a foot, the bird dismembers it with its bill. If an escaping insect drops to the ground, the automolus follows it and hops about, flicking the litter aside with its bill in an effort to retrieve it. Then it sidles up slender stems or vines with frequent about-faces, nervously twitching its reddish brown wings and voicing harsh notes, to resume its rummaging in a tangle of vines or to dart beyond view. Occasionally an automolus briefly joins a motley assemblage of small birds catching fugitive insects stirred by army ants (not the ants themselves); but no ovenbird is known to be a regular or "professional" ant-follower.

Another rummager is the Streaked-breasted Treehunter. This northernmost representative of an Andean genus inhabits midaltitude forests in Costa Rica and western Panama. Here it is found in dense stands of small trees and bushes, often in deep ravines along mountain streams, where mosses and larger epiphytes grow lushly. In this excessively humid milieu, the treehunter ransacks mossy limbs and the dead leaves caught up among them for insects, spiders, frogs, salamanders, lizards, and other creatures. One day I watched a treehunter engaged at a large, epiphytic tank bromeliad, the elongated leaves of which, arranged in an open rosette, held much rainwater between their tightly overlapping bases, forming a little aerial pond that supported a varied fauna. The stout-billed bird tore and tugged at the narrow leaves, pulling off large pieces and dropping them to the ground. After I had watched, greatly puzzled, for about five minutes, the bird revealed the purpose of its strenuous activity by extracting from among the bases

Striped Treehunter *Thripadectes holostictus* Sexes similar Venezuela to Bolivia

of the leaves a small frog, which it promptly carried to the ground, doubtless to devour it. Other rummagers are Striped, Lineated, and Spectacled "foliage-gleaners," three species that probe into clumps of mosses and other epiphytes and search accumulations of trash in vines, branches, and tree-fern crowns in humid forests at low and middle altitudes. In northern Bolivia, J. V. Remsen, Jr. and T. A. Parker, III watched nine species of ovenbirds who spent over 75 percent of their foraging time hunting insects in dead, curled leaves suspended above ground in plants.

Foliage-gleaners, properly so called, seek insects and spiders on living leaves and the branches that bear them. The Buff-fronted Foliage-gleaner, a slender brownish bird of montane forests in Costa Rica, finds much of its food at the ends of twigs and shrubs, high in the forest or low at its edges. It forages much like a vireo but is more active, rarely pausing, frequently clinging head-downward to pluck a caterpillar from beneath a leaf. Usually I have found the bird silent and alone.

In northern South America, the Cinnamon-rumped Foliage-gleaner clambers over the fronds of palms and the large leaves of heliconias, aroids, and other plants, or hangs from their tips, in the open understory of forests or, rarely, up to the subcanopy. In Colombia, Double-banded Graytails forage much like Buff-fronted Foliage-gleaners. In some of its diverse habitats, the Plain-mantled Tit-Spinetail, widespread in southern and western South America, gleans foliage in light woods and brushland, often hanging head-downward to explore the undersides of leaves and twigs, much in the manner of the Eurasian Long-tailed Tit for which it is named. The Thorn-tailed Rayadito gleans from foliage and trunks.

Twig-gleaners include the Plain Xenops and its relatives. One needs only to look closely at the upcurved lower mandible of a xenops's short bill to suspect that it forages in a unique way. It finds much of its food in slender dead twigs, especially those, far advanced in decay, that have broken from the parent tree and are caught up in vines that drape the tree, or are clasped by tendrils that prevent their fall. The tiny bird climbs over and clings to these branches in all attitudes, erect

Plain Xenops *Xenops minutus* Sexes similar Southern Mexico to northern Argentina

White-throated Treerunner *Pygarrhichas albogularis* Sexes similar Southern Argentina and Chile to Tierra del Fuego

Wrenlike Rushbird *Phleocryptes melanops* Sexes similar Southern South America to Tierra del Fuego

or inverted, while it hammers on them with a wedgelike bill, putting its whole body into each blow, much in the manner of a chickadee. After it has opened a gap it may insert its bill and push upward, breaking away the soft wood with the sharp ridge of its upper mandible, using it like a can-opener, until it overtakes the larva that has retreated upward through the pith. The upcurved lower mandible makes a bill at once sharp for pecking into wood and reinforced for exerting the upwardly directed pressure needed for elongating the gap that it has made.

Like the piculet that it resembles, the Plain Xenops is fond of ants. I watched one clinging to the thick petiole of a large dead leaf of a cecropia tree that had lodged on a lower branch of an adjacent tree. The brown bird pecked vigorously at the dry petiole until small ants swarmed out of the pithy center, then snatched up and swallowed all that it could catch. Another xenops clung to a thick dead vine and enlarged a gap in the bark, through which it extracted and devoured many white ant pupae, and probably also the mature ants that attended them.

Still another xenops was pecking into a small dead branch near its end. After hammering for a while, it would go to the broken-off end and peer up into the hollow with one eye, an act characteristic of the bird, who thereby keeps track of its prospective victim. After each inspection, the bird resumed its pecking, until it drew a big brown insect from the branch. In much the same way, it removes long, slender larvae from the centers of dead vines. If the prey drops out of the end of a stem, the xenops follows it to the ground, where otherwise I have not seen it. The bird also gives attention to shriveled dead leaves attached to the ends of slender twigs, often hanging head-downward to explore their folds. Other species of *Xenops* appear to forage in much the same way. The Great Xenops, of a different genus, employs its wedgelike bill to lift up loose bark on middle-sized branches over which it clambers, to extract small insects beneath it, often without detaching the scales of bark.

A trunk-gleaner is the White-throated Treerunner, most woodpeckerlike of ovenbirds, which in the forests of southern Chile and western Argentina climbs up trees and over thick branches, using its spine-tipped tail feathers for support, while it gathers insects from the bark or exposed wood of dead trees. With the establishment of plantations of pine trees (absent from South America except where introduced), it has been extending its range northward. When it lives in forests, the little Thorn-tailed Rayadito runs swiftly up and down the trunks of trees with its queer tail tilted up and its body teetering up and down, while it searches for insects in crevices in the bark. In more open habitats, such as the islands of the Cape Horn Archipelago, the rayadito is a shrub-and-ground-forager, searching amid fallen leaves in company with Rufous-collared Sparrows and other finches.

To complete our survey of the very diverse modes of foraging by insectivorous ovenbirds, we turn to the marsh-dwellers. The Wrenlike Rushbird forages amid the reeds and cattails of its marshes, or occasionally ventures forth into adjoining patches of grass and other herbs. Where large-leaved floating vegetation covers the water, the small birds hop around on it, frequently with their feet and legs immersed in the cold water, while they eagerly seize small aquatic creatures. Where such support is absent, they cling to vertical stalks and stretch down as far as they can to probe with their bills the small floating plants, like duckweed, that cover the surface. Two other marsh-dwellers of southern South America appear to find most of their food amid reeds and rushes.

From seacoasts through deserts, scrublands, marshes, and rain forests, up to the highest Andean slopes where hardy plants grow, adaptable ovenbirds have found ways to nourish themselves, all without deviating widely from their predominantly insectivorous diets.

14 *Daily Life*

Much of a bird's day is occupied by the indispensable business of finding food, but because foraging is such a large subject, it has received a separate chapter. In this chapter we consider other aspects of ovenbirds' lives, beginning with their social relations. Many ovenbirds are found in pairs at all seasons, a fact that has led naturalists to conclude too hastily that they mate for life. Living always in pairs is obviously not incompatible with frequent changes of partners, although among birds this is improbable. Only by following individually recognizable birds for years can long fidelity be proved. In a seven-year study of a partly banded population of Rufous Horneros in Argentina, Rosendo Fraga found two pairs who remained together throughout the year, including four consecutive breeding seasons, and one pair for three seasons; these horneros evidently formed lifelong bonds. Another male changed mates shortly after the end of each breeding season, although at least two of his partners were still alive when he separated from them. In every case, this fickle male obtained a new spouse in less than a month, with the result that he was almost constantly paired in spite of nuptial inconstancy.

Four pairs of the Brown Cacholotes watched by Ana and Manuel Nores remained intact throughout the four years of their study, eight pairs for three years, and thirteen pairs for one year. Two banded pairs changed partners during a breeding season. When a mated cacholote died, the bereaved bird of either sex promptly found a new spouse. A banded pair of Rufous-fronted Thornbirds remained intact through the four years that Betsy Thomas followed them. Among ovenbirds that are found in pairs between breeding seasons and apparently mate for life are Common Miners, Slaty Spinetails, White-whiskered Spinetails, Firewood-gatherers, Ruddy Treerunners, White-throated Cacholotes, and Curved-billed Reedhaunters.

Ovenbirds that I have usually found singly after their young have become independent are Streaked-breasted Treehunters, Red-faced Spinetails, and Buff-fronted Foliage-gleaners. Solitary foraging is certainly not proof that pairs have been dissolved; the partners may merely find it more profitable to hunt apart. Monogamy is the rule among ovenbirds. Every nest of which I have sufficient information was attended by both parents, except one of the Buffy Tuftedcheek which may have been abnormal.

Whether in pairs or solitary, ovenbirds rarely join in flocks of their own species. Those that do not migrate appear to remain on their territories throughout the year, as Fraga proved of his banded Rufous Horneros. Both sexes of these horneros defended territories of about 0.6 to 2.5 acres (0.23 to 1 hectare) by chasing or fighting, with much calling. Except when new pairs tried to win territories in occupied areas, conflicts were brief. These were in Buenos Aires Province, beyond the tropics. Within them, I have never seen ovenbirds of any species fight, and rarely passerines of other families.

The most gregarious ovenbirds live in the far south, where winters are harsh. In Patagonia in the cold season, W. H. Hudson found Plain-mantled Tit-Spinetails in flocks of thirty or forty individuals, with ovenbirds of other species, mainly the Lesser Canastero, all advancing through wintry thickets together, carefully exploring every bush. Also in Patagonia after the breeding season, François Vuilleumier watched Thorn-tailed Rayaditos foraging in the beech forests in flocks of four to fifteen birds, frequently accompanied by one or more White-throated Treerunners and more rarely a little Striped Woodpecker, all moving rapidly and incessantly in search of food. Another Patagonian bird, the partly migratory Scaly-throated Earthcreeper, sometimes unites in small, loose flocks. Of another partly migratory ovenbird, the Bar-winged Cinclodes, Hudson wrote: "These birds cannot be called strictly gregarious, but where abundant they are fond of gathering in loose flocks, sometimes numbering one or two hundred individuals, and when thus associating are very playful, frequently pursuing and wheeling about each other, and uttering a sharp, trilling note."

In tropical forests ovenbirds forage alone, in pairs or family groups, but not, as far as I have seen, in flocks of one species. However, a single individual or a pair of ovenbirds often join a mixed-species flock, which in rich woodland may become large, with many individuals of a dozen or more species of different families. Participation in these wandering flocks is not incompatible with strict territoriality. As the flock crosses the territorial boundary of one of its members, this individual or pair drop out, to be replaced by an individual or pair of the same species on the other side of the unmarked line of separation. Sometimes neighbors display in an assertion of possession as the flock crosses the boundary, but often the substitution is so quietly effected that, unless the birds are individually recognizable, as by banding, one receives the impression that the same individual accompanies the flock throughout its wanderings, ignoring territorial limits.

Probably ovenbirds bathe as frequently as other kinds of birds, but I have seen this only thrice, and have found no published record of this activity. In a little rill flowing through a Panamanian forest was a pool where a pair of Scaly-throated Leaftossers bathed. Almost every evening in May when I passed the pool after six o'clock, I found them splashing long and vigorously in the shallow water. Their ablutions over, they would fly through the undergrowth to sleep where I could not find them. If the leaftossers noticed me watching them, they called sharply.

One day in August, I saw a Plain Xenops slip into a cavity with a wide opening, about twenty feet (6 m) up in a tree trunk. Here it stayed for a minute or two, looking out, sometimes backing farther into the hole and sometimes exposing more of its body. From the fine drops that flew through the opening, and from the matted appearance of the bird's plumage when it emerged, it was evident that it bathed in rainwater that had collected in the cavity. Late in the afternoon, after a hard rain, a xenops came again to the pool and bathed in the same manner.

I have repeatedly watched a Plain Xenops engage in the widespread, puzzling avian activity called "anting." In a typical episode, the xenops clung in an upright posture to a slender twig beside a silken nest of the small, stingless black ant *Camponotus senex,* attached to a hanging dead branch high in a tree at the woodland's edge. The bird appeared to pluck something from the surface of the ants' nest or from a twig close beside it. Then, bringing a wing forward in front of its body, it ran its bill rapidly over the lower or inner surface of the remiges. As it did this, its tail was bent forward beneath its nearly upright body. The bird repeated this act about a dozen times in rapid succession. Probably because of its height, I could detect nothing in its bill, but analogy with other performances of this sort left little doubt that the xenops held an ant each time it rubbed its bill over its wing feathers. Whether after this it ate the ant, as often happens in such cases, I could not tell. Although in the North Temperate Zone birds of many kinds usually ant on the ground, in the tropics they nearly always perform in trees or shrubs, not excluding terrestrial birds. Whether birds ant to combat feather lice with formic acid, to prepare prey for swallowing, or simply for the sensation, is not well known.

Probably all birds preen themselves, and many preen their mates or other close associates, but I have only one record of allopreening in ovenbirds. On a morning in July, harsh rattles drew my attention to a pair of Slaty Spinetails amid bushy growth. Presently one perched in a small tree with its head bent down and inward until its bill almost touched its breast, with the feathers of its head and neck all standing out. Its mate came and perched beside it to nibble at the outfluffed plumage. After performing this service for a while, the preener picked up a stick from a small pile of them in a nearby tree, possibly the beginning of a nest. The first spinetail remained in the same posture, as though inviting more attention, but its partner preferred to move sticks.

In western Ecuador, I saw a Pale-legged Hornero give a morsel to its mate, with whom it joined in shrill duets. In Costa Rica, I watched two Plain Xenops cling side by side while they gathered insects, probably ants, that swarmed out of a dead vine. One repeatedly passed some of the insects to the other, who appeared equally mature and was probably its mate rather than a dependent young bird. In Venezuela, Betsy Thomas saw just one instance of nuptial feeding in pairs of Rufous-fronted Thornbirds that she followed for periods up to four years. In five months of intensive study of thornbirds in the same country, I did not see an adult feed its mate. Nuptial feeding appears to be rare in the ovenbird family.

As day ends, many ovenbirds go to rest in holes, burrows, crannies, or nests as diverse as those in which they rear their families. Miners sleep in burrows that they dig. Bar-winged Cinclodes retire at nightfall into burrows, holes in dams, crannies beneath the roofs of buildings, under bridges, in dense hedges, or similar sheltered refuges. Streaked Tuftedcheeks lodge in holes in stream banks or under tufts of grass on the páramos, where trees are absent. Probably at lower altitudes they sleep in cavities in trees where tuftedcheeks nest. On the puna of Peru, Oliver Pearson found immature female Andean Tit-Spinetails passing the night in caves and mine tunnels, where hummingbirds and larger birds also slept, all seeking shelters that remain considerably warmer than the open puna, which on clear nights can become piercingly cold. Whether miners, cinclodes, tuftedcheeks, and tit-spinetails sleep singly or with companions, my sources do not reveal.

Plain Xenops pass their nights singly in holes in trees, made by themselves, by woodpeckers, or by decay. High in a dead avocado tree in a small coffee plantation, one lodged in a cavity carved and abandoned by a pair of Golden-naped Woodpeckers, who now slept together in another branch of the same tree. The contrasts between these neighbors were interesting. The larger woodpeckers retired much earlier in the evening, and remained inside many minutes longer in the morning than the xenops did. Like other woodpeckers, they cautiously inspected their dormitory before they entered, making sure that no unwelcome intruder had preceded them. In the fading light when it was difficult to see, the xenops flew up from a distance and, without pausing, shot right in, with no preliminary investigation. In the dim dawn, I needed to watch carefully not to miss the tiny bird's abrupt departure. Much shier than the woodpeckers, if it noticed me watching in the evening, it flew off, probably to some other lodging that it knew. But after the bird had entered its hole, neither tapping nor scraping on the trunk would make it show its head in its high doorway.

One evening in March, when the xenops arrived at its hole, it did not, as usual, enter unhesitatingly, but clung for several seconds in front of the orifice. Soon another xenops, that I had not previously noticed, emerged from the hole, and the two flew swiftly to nearby trees, as though one pursued the other. After a while, one returned and entered the hole, so rapidly that I hardly saw it. Soon the other bird returned, and while it clung beside the doorway the first emerged and flew off through the dusk. The latest arrival went in, and from the interior I heard the xenops's fine trill repeated several times. On other evenings, the bird was always silent after entering. A few days later, a second xenops again tried to join the first in the cavity, but only one remained. It was clear that xenops prefer to sleep alone. At this season, the two birds who coveted the same cavity were probably mates,

Red-faced Spinetail *Cranioleuca erythrops* Sexes similar Costa Rica to western Ecuador

but I could not distinguish their sexes. The Thorn-tailed Rayadito also sleeps in a hole in a tree, but whether singly or with others I do not know.

In a coffee plantation beside forest in the highlands of southern Costa Rica, a Red-faced Spinetail slept alone in a large, globular nest, composed mainly of green moss and thin vines, that hung about thirty feet (9 m) up at the end of a drooping leafy branch, where I could not reach it. In the evening, a solitary spinetail arrived silently through the treetops and darted into an inconspicuous doorway in the nest's side, near the bottom. It entered so swiftly that unless I kept sharp watch I would miss it. When the dawn light was still dim beneath the trees, the bird shot as suddenly and swiftly out and away. On the rough, vertical face of a cliff beside a highway in Venezuela's Coastal Range, Paul Schwartz showed me inverted mossy pockets, less bulky versions of the Spotted Barbtail's unique nest (described in chapter 16), in which these brown-and-olive ovenbirds slept singly throughout the year.

At no great height in trees and bushes of central South America, Striped-crowned Spinetails build for sleeping tubular nests of twigs, vegetable fibers, tendrils of clematis, and wool. At each end, near the bottom, is a doorway. If a nocturnal prowler pokes into one of these openings, the sleeper might escape through the other. It is said that to

Hanging globular nest of Red-faced Spinetail in Costa Rican highlands

convert the dormitory into a nest for eggs and young, the spinetails close one of the doorways, but this seems unlikely because breeding nests have a different form, being globular with a single lateral entrance near the top. Another bird that sleeps in a covered nest with two doorways and breeds in a structure with only one is the White-browed Sparrow Weaver of Africa.

In Argentina, several Larklike Brushrunners help to build in a tree or shrub a large globular nest of interlaced sticks, with the entrance at the top. Up to eight individuals have been found sleeping in such nests. Parents and juvenile Firewood-gatherers sleep together in their houses of sticks, according to Hudson for three or four months after the young fledge, but Fraga found that the juveniles were expelled less than one month after they flew. The Nores's Brown Cacholotes slept throughout the year in their spacious nests of sticks, but only for five to fifty-five days (average forty-two days) in any one of them. Juveniles remained in the parental territory, at first sleeping in the nest

Larklike Brushrunner *Coryphistera alaudina* Sexes similar Bolivia and southern Brazil to Argentina

where they were reared, but soon in their parents' new dormitory. The Rufous-fronted Thornbirds' many-chambered nest serves as a family residence throughout the year, as will be told in chapter 19.

Although our knowledge of how ovenbirds sleep, as of the sleeping of most other birds of tropical America, is fragmentary, those known to pass the night under a roof of some sort are so diverse in habits, and so widely scattered geographically and taxonomically, that it appears probable that most members of the family occupy dormitories instead of roosting amid foliage. However, exceptions occur. Rufous Horneros do not sleep in their ovens, where they might be trapped by a predator blocking the doorway. Fraga found them roosting in dense foliage, which in winter was available only in evergreen trees or shrubs. If a hornero had no suitable tree in its own territory, it entered a neighboring territory with better accommodations, surreptitiously, after the territory owners had retired—behavior not unlike that of Rufous-fronted Thornbirds who have lost their nests.

Although I have watched many nests of Rufous-breasted and Slaty spinetails, I have never found a grown bird in any of them at night, unless it was incubating eggs or brooding nestlings. In addition to this negative evidence, I have a bit of positive evidence about the way they sleep. One January, a pair of Slaty Spinetails roosted in the tall molasses grass and pokeberry bushes that densely covered an old maize field beside forest. While studying birds in this clearing, I often heard the spinetails as they awoke at dawn. The first to wake up called with a throaty rattle and was promptly answered by its mate, who, to judge by the sound, slept a short distance away. In the evening, I sometimes heard the pair communicating as they went to rest. I searched in vain for a nest in which they might have slept. In January, at the end of the long rainy season, nests made for breeding during the preceding year are too dilapidated to be habitable, and the construction of the earliest new ones has just begun. I believe that Slaty and Rufous-breasted spinetails, like Rufous Horneros, depart from the widespread ovenbird habit of sleeping in dormitories and roost in low, dense vegetation.

15 *Voice, Displays, & Courtship*

With simpler vocal organs than songbirds (oscines) have, ovenbirds cannot produce the elaborate musical figures that delight us in the songs of such accomplished musicians as thrushes and wrens. Their longer utterances tend to be repetitions of a single note or pair of notes, as in trills, twitters, chips, churrs, or rattles. They achieve variety by accelerating or retarding their tempo, increasing or reducing the volume. Often the trill or churr is introduced or terminated by one or a few different, emphasized notes. Typical examples are the *chi-chi-chi-chi-ti-ti-ti-'t 't* of the Silvery-throated Spinetail in falling cadence, the slightly ascending *tetet't't't't't't't't* of the Bar-winged Cinclodes, and the *chee-chee-chee-chee-chuchu chu chuchee-chee* of the Blackish Cinclodes. The Wrenlike Rushbird's peculiar song consists of a long, cicadalike buzz, followed by a series of sounds like small taps on a piece of dry wood. Another strange song is that of the Red-faced Spinetail: a rapid flow of sharp notes in a tone that seems forced and artificial begins at a high pitch and ascends to thin notes that appear to be at the upper limit of human audition.

Ovenbirds' recitals can become extremely monotonous. Hudson, who knew the Pale-breasted Spinetail in Argentina, near the southern extremity of its vast range, wrote: "I am pretty sure that in Buenos Aires it is migratory, and as soon as it appears in spring it announces its arrival by its harsh, persistent, two-syllabled call, wonderfully strong for so small a bird, and which it repeats at intervals of two or three seconds for half an hour without intermission. When close at hand it is quite as distressing as the grating song of a Cicada. This painful noise is uttered while the bird sits concealed amid the foliage of a tree." More than a century later, in southern Costa Rica at the

Pale-breasted Spinetail *Synallaxis albescens* Sexes similar Costa Rica to central Argentina

opposite end of this spinetail's far-flung range, I heard it perform much as Hudson described. Its call is a sharp, emphatic *bet chú,* and its song, if it may be credited with one, is a sequence of the same dry notes: *bet chú bet chú bet chú* . . . One bird delivered 36 to 41 *bet chú*s per minute for thirty-five successive minutes, during which it uttered about 1,300 of these disyllables. Not infrequently this spinetail, who usually lurks low amid coarse grasses and herbs, rises to a fairly high, exposed branch to recite. I watched one who called for a long time from twenty to thirty feet (6 to 9 m) up in the open crown of a roadside tree. With each repetition of the *bet chú,* the feathers of the bird's throat stood out momentarily, revealing their dusky bases and making a fugitive dark patch where most of the time none is seen. How conservative are birds, to preserve the same manner of singing, as of building their nests, during the many generations that must separate populations forty-five degrees of latitude apart!

Another ovenbird that sings very persistently without achieving melody is the Buff-throated Automolus. Its most frequent utterance is a loud, harsh, prolonged rattle, which during much of the year sounds in the forests of Central America in the early dawn when these birds first awake, and again as the shadows of night descend through clouds of foliage. The length and tempo of these recitals vary greatly. With his tail beating time to his rapid notes and his throat puffed so far out that he appears to wear a fluffy beard, the automolus delivers songs of seven or eight notes, each song lasting two to five seconds. One bird sang at the rate of about eleven songs per minute. Before sunrise on an

April morning, an automolus began his recital by pouring out twenty-one songs per minute. At times the female answers her partner with fewer and shorter rattles.

At dawn in the nesting season, I have heard the Plain Xenops sing no less persistently than his larger neighbor, the Buff-throated Automolus, but in a very different voice. His song is a fine, sharp, very rapid trill, so similar to that of his counterpart the Olivaceous Piculet that it is difficult to distinguish them; but that of the piculet is softer. The first note of the xenops's trill is emphasized. One who was already singing when I came within hearing continued to repeat his song at irregular intervals for the next twenty-five minutes. The concentrated dawn singing of ovenbirds corresponds to that of American flycatchers, another family of birds with limited vocal powers that they display chiefly in the dim light before sunrise.

The simple songs of ovenbirds can be appealing and occasionally beautiful. As heard by Hudson, the Common Miner "trills out its clear, ringing rapidly reiterated cry, which in sound resembles the laughter of a child." My favorite of all the ovenbirds' songs that I have heard owes its charm not to variety but to the quality of its clear, ringing notes. The Scaly-throated Leaftosser's song consists of two ascending trills, the second of which ends in a little silvery tinkle, or sometimes with dry, chaffy notes; the clearest notes come in the middle of a sequence. Each verse, of a dozen or more notes, lasts five or six seconds. When the bird performs most intensely, the high notes that terminate one verse are separated from the full, deep notes that introduce the next verse by a scarcely perceptible interval. A bout of singing often continues for many minutes, but despite its repetitions it is too beautiful to become tiresome.

Ovenbirds have been reported to sing in flight less often than one expects of birds that frequently inhabit country where trees are few or lacking. While flying swiftly and low, Scaly-throated Earthcreepers voice shrill trills that resemble laughter. On warm winter days, Hudson watched Bar-winged Cinclodes rise vertically into the air to pour out with great energy a confused torrent of unmusical sounds. While flying on a slightly undulatory course, the Blackish Cinclodes utters snatches of its twittering song.

Dueting or responsive singing is widespread among ovenbirds. For a vivid description of the duets of Rufous Horneros we turn again to Hudson, who grew up among these birds.

> At frequent intervals during the day the male and female meet and express their joy in clear, resonant notes sung in concert. . . . The first bird, on the appearance of its mate flying to the place of meeting, emits loud, measured notes, sometimes a continuous trilling note with a somewhat hollow, metallic sound; but immediately on the other bird joining, this introductory passage is changed to rapid triplets, strongly accented on the first and last notes, while the second bird utters a series of loud measured notes perfectly according with the triplets of the first. While thus singing they stand facing each other, their necks outstretched, wings hanging, and tails spread, the first bird trembling with its rapid utterances, the second beating on the branch with its wings. The finale consists of three or four notes uttered by the second bird alone, and becoming successively louder and more piercing until the end. There is infinite variety in the tone in which different couples sing, also in the order in which the different notes are uttered, and even the same couple do not repeat their duet in precisely the same way; but it is always rhythmical and to some extent an harmonious performance, as the voices have a ringing, joyous character.

Rufous Horneros begin to duet at an early age. During their last week in the nest, their voices are audible at a considerable distance. When about to leave the nest they frequently answer parental calls and may even attempt to duet with a parent, as Rosendo Fraga learned.

When, after a brief separation, a male and female Firewood-gatherer meet at their nest, they sing in concert a verse composed of four or five loud ticking chirps, followed by a long trill, "as if rejoicing over their safe reunion," as Hudson wrote. When a pair of Brown Cacholotes meet, their loud screaming duet rings through the woods. Families of White-throated Cacholotes join in a piercing chorus that sounds like screams of hysterical laughter, audible perhaps a mile away. High above tree line in the Andes, Streaked-backed Canasteros sing responsively. As they forage through dense vegetation where they cannot see each other, mated Slaty Spinetails keep in touch with the rattling *churr* that seems to be their only song. I have already mentioned how the female Buff-throated Automolus answers her mate's prolonged rattle. While rummaging amid the densest vine tangles, Ruddy Automoluses communicate with throaty rattling notes, and at intervals voice a loud *knaayr knaayr,* between a whine and a scream—the

Many-striped Canastero *Asthenes flammulata* Sexes similar Colombia to Peru

strangest, most unbirdlike sound that I have heard from an ovenbird. Of the Rufous-fronted Thornbirds' duets I shall tell in chapter 19.

In addition to longer recitals, ovenbirds have calls that often sound much like the notes which, strung together in longer sequences, compose their songs. A voluble species that I knew well without ever hearing a sustained performance that I was inclined to call a song is the Rufous-breasted Spinetail. Often, floating out of a tangled thicket, I heard its pleasantly conversational *pet chu, pet pet pet pét chu,* delivered with rising inflection and emphasis on the last *pet.* The final *chu* was shorter and softer. More rarely I heard a rolling *uddle uddle uddle uddle,* the nearest to a song that I noticed. A longer utterance, sometimes heard from Rufous-breasted Spinetails attending nests, sounded like *krr-r krr-r; krr kr kr, witta witta witta witta witta witta wit,* the first phrase pronounced slowly and deliberately, the second accelerated. This call, of which I noticed several versions, lacked melody but was engaging. Issuing from thickets and canefields on the slopes of Volcán Tungurahua in Ecuador, I frequently heard the dry, good-natured *peck weck, peck weck* of Azara's Spinetails that rarely gave me fleeting glimpses of themselves.

Among constantly paired passerine birds, nuptial displays tend to be inconspicuous, rare, or absent. A Pale-legged Hornero, crouching on a branch beside an unfinished nest with its mate nearby, held its spread wings motionless, displaying their alternating pale yellow and dark bands, then closed them slowly, singing the while. Displaying Dark-breasted Spinetails puff out their white throat feathers to reveal their bases as a patch of solid black. Neither of these displays has been seen often enough to reveal its significance.

The Nores witnessed courtship by the Brown Cacholote only near or inside the nest, where they watched it in structures with thinner walls. After repeatedly joining in a duet with his consort, the male, with body almost horizontal and plumage bristling, raises and spreads his tail, droops his wings until they almost touch the ground, depresses his bill, and voices a series of short notes at the rate of about two per second. In this posture, he enters the spacious chamber of the covered nest. The watching female follows him but stays in the entrance tunnel, while he rotates two or three times in the chamber, stopping with his bill directed toward her. Then the female joins him and crouches low for about fifteen seconds. With feathers still erect, he approaches and circles around her, then mounts her with depressed tail while she raises hers. After he dismounts, they duet inside the nest.

While watching a Creamy-breasted Canasteros' nest amid the massive leaves of a giant terrestrial bromeliad (*Puya raimondii*) on the high puna of Peru, Jean Dorst became the center of an unusual mobbing display. When the parents returned with food for their nestlings and found him sitting nearby, they started to skip, rapidly and with great agility, over the long, spine-bordered leaves. They raised and lowered their closed tails and loudly repeated a metallic roll: *brreu brreee . . . brreu*. This dance promptly attracted others of their kind, who from all sides flew up alone or in pairs, until about fifteen were jumping over the puya, pumping their tails and calling excitedly. The active crowd dissolved only when the watcher departed. Although these canasteros are often shy in the presence of humans, when their young appeared to be threatened they became bolder, attracted their neighbors with a special harsh cry, and in a manner unexpected in territorial birds, permitted their close approach to the nest while all together tried to distract a man who seemed to threaten it.

A. W. Johnson watched a dozen Thorn-tailed Rayaditos mob an owl with a deafening chorus of scolding notes. When a human visits a nest, the parent rayaditos flit from branch to branch around him, almost touching his head, complaining or scolding loudly in a typical mobbing display.

I have never seen any ovenbird try to lure an intruder from its nest or fledglings by fluttering or limping away as though injured and unable to fly, and I have found no published description of such "injury feigning" in the family. These distraction displays are more frequent in birds with open nests, such as antbirds and wood warblers, than in birds which breed in enclosed nests, holes, or burrows. The safety of these nests appears to depend more upon escaping the notice of predators, or inaccessibility to them, than upon enticing them away.

16 *Nests*

Nests of ovenbirds, often very much larger and easier to find than the small birds who make them, have attracted more attention than any other aspect of these birds' lives. In form they range from burrows in the ground to mansions of interlaced sticks, sometimes high in trees. The only feature that all have in common is their provision of an enclosed, roofed space for eggs, nestlings, and attendant parents; no ovenbird is known to build an open cup amid foliage, such as most passerines use. How to survey such a great variety of nests without tiresome prolixity is puzzling. It seems best to treat them in the order of increasing complexity, although this will entail some inconsistencies, and the separation of closely related species, for those in the same genus sometimes build quite different nests.

Let us begin with ovenbirds that nest in burrows in the ground, whether in treeless regions or tropical rain forests. Among the former is the Common Miner, which Hudson called the Little Housekeeper. On the Argentine pampas it is attracted to colonies of the Vizcacha, a large rodent, in the vertical sides of whose deep burrows it bores tunnels three to six feet (1 to 2 m) long, terminating in a round chamber that it lines with soft, dry grasses. In Argentina and Chile other naturalists have found Common Miners' burrows from twenty inches to twelve feet (0.5 to 3.5 m) long, in sand dunes and vertical or sloping exposures of earth, and loosely lined with cardoon inflorescences, straws, rootlets, rodent hairs, or feathers, forming an open cup for the eggs. Other miners and earthcreepers occupy similar burrows, which they often dig for themselves. An exception is the Chaco Earthcreeper, which in fissures amid rocks, crannies in buildings, cavities in trees, and old nests of Rufous Horneros arranges for its eggs a loose bed of

Scaly-throated Leaftosser *Sclerurus guatemalensis* Sexes similar Southern Mexico to Colombia

such diverse materials as straws, rodent hairs, snakeskin, fragments of newspaper, aluminum foil, and plastics.

On high, rocky slopes of the front range of the Chilean Andes, Crag Chilias excavate a burrow or choose a hollow in a cliff or cactus for their nests loosely built of dry twiglets and stiff feathers. The eight species of cinclodes select nest sites as diverse as their habitats from the high Andean páramo and puna to the seacoast. Often they dig burrows in exposures of soft soil, preferably near water, but frequently they occupy crevices in cliffs or stone walls, holes in trees, old kingfishers' burrows, rodent burrows, nooks in buildings or bridges, and the like. Into the chosen site they carry grasses, feathers, and other soft materials for a loose but bulky nest.

Among the ovenbirds that nest in burrows amid tropical forests are several species of leaftossers. They dig their tunnels in streamside or roadside banks, or in the vertical wall of rootbound soil, sometimes more than head-high, heaved up by the fall of a great, shallow-rooted tree. The burrows, from twelve to thirty-two inches (30 to 81 cm) long, are wider than high at the mouth. They usually curve far enough to the right or the left to prevent a view of their contents from the front without a small mirror and a flashlight bulb on a long cord. I watched a pair of Scaly-throated Leaftossers excavate a burrow in a root mass where four successive broods were raised. With much singing, both sexes dug, beginning well in advance of laying and proceeding slowly. Sometimes they were together in the burrow. Instead of throwing loosened earth backward with vigorous alternating kicks with their feet each time they entered, as jacamars, motmots, and kingfishers do, they came to the entrance and flicked it out with muddy bills. In the expanded inner end of their burrow the leaftossers arranged a loose, shallow nest of the rachises of compound leaves from which the leaflets had fallen. When the eggs were laid, this mat was sometimes so thin that the eggs touched the ground, but the birds continued to add pieces to the nest during incubation.

In the same forests where Scaly-throated Leaftossers dwell, Buff-throated Automoluses dig burrows in more or less vertical banks beside streams or roads, or in the side of a pit. Excavation may start as early as October, when the earth is moist and soft, although the burrow will not contain eggs until the following March, when the soil is dry and hard. Another advantage of digging a burrow so long in advance of occupancy is that the pile of excavated earth below its mouth is no longer so fresh and conspicuous that it might attract predators. (Blue-diademed Motmots, who sometimes nest in the same banks with automoluses, prepare their burrows with similar foresight.) Occupied automoluses' burrows that I have seen were usually twenty-four to twenty-nine inches (61 to 74 cm) long, but one measured only eighteen inches (46 cm). In the chamber at the inner end the birds arrange a broad, shallow bowl of a single kind of vegetable material, usually curving, leafless brown rachises of compound leaves. In wet, midelevation forests of Costa Rica and western Panama, Streaked-breasted Treehunters make a bulky cup of interlaced, fibrous rootlets in the chamber at the end of a burrow about two feet (61 cm) long. Other burrow-nesters in tropical forests are certain foliage-gleaners of the genus *Philydor,* but little information about them is available.

Tit-spinetails of the genus *Leptasthenura* sometimes occupy burrows that they did not dig, but perhaps more frequently they choose closed spaces of other kinds, including often old ovens of Rufous Horneros, stick nests of canasteros, holes in trees or cacti, nooks in buildings, or almost any other cavity that offers shelter for a loosely built nest of twiglets and rootlets, so profusely lined with feathers that it becomes the snuggest of nests. If the nest chamber is entered through a passageway or tunnel, tit-spinetails line this, too, with feathers and other soft materials, often leaving feathers protruding from the doorway, incautiously advertising that their nook is inhabited. In the forests of southern Chile and neighboring Argentina, at their edges or in nearby clearings, the Thorn-tailed Rayadito nests beneath loose flakes of bark, in clefts in trunks, or other crannies. On a foundation of rootlets and straws, it spreads a soft bed of feathers for its eggs.

Other cavity-nesters are several species of xenops. In a decaying, upright trunk or branch, in or near woodland, Plain Xenops carve a neat miniature of a woodpeckers' hole, with a round doorway slightly less than an inch (2.5 cm) in diameter. The sexes alternate in pecking into the soft wood, much as a woodpecker does, sometimes working continuously for nearly an hour. They line the cavity with soft bast fibers, which they continue to bring during incubation. On one occasion a pair of Plain Xenops had eggs in a low hole carved by Olivaceous Piculets, so much like their own work that I would not have suspected its different origin if I had not earlier found piculets sleeping in it while it was still unlined. Their only alteration was the addition of a soft lining, which piculets do not use. Most woodpeckerlike of all ovenbirds is the White-throated Treerunner, which in forests of southern Chile and across the Andes in Argentina carves in rotting stumps or trunks or half-burned trees well-rounded cavities up to one foot (30 cm) deep, such as a woodpecker might make. A bed of wood fragments mixed with a few straws covers the bottom where the eggs are laid.

The nest of the Streamside Sharptail (also called Sharp-tailed Streamcreeper) contains features of the foregoing birds' nests and of those we have next to consider. At the expanded end of a burrow up to two feet (60 cm) long, in at least one case left by the decay of a thick root in a streamside bank, the sharptail fashions a globular nest of rootlets and twiglets, lined with bamboo leaves. It has a side entrance and might be compared to an Ovenbird's nest on the ground in a northern woodland. Other sharptails' nests are built in crannies amid rocks. Why the sharptail makes a covered nest deep in a burrow is not clear; most burrow-nesters dispense with this luxury. Possibly the nest is roofed to shield its occupants from falling bits of earth, but I surmise that the covered structure is an inheritance from ancestors that built such nests in the open. Another ovenbird that breeds in tunnels and other closed spaces, where it often but not always makes a globular nest with a side entrance, is the Cordilleran Canastero, most of whose closest relations build enclosed nests in the open, as the ancestors of the Cordilleran Canastero probably did.

Globular nests opening sideward, similar in form to that of the Streamside Sharptail, are built on the ground by several unrelated ovenbirds. The nest of the Córdoba Canastero is situated in a compact tuft of grass, stalks of which are built into its walls, which are completed with fine grasses brought by the builders, with sometimes the addition of lichens and mosses. Often the doorway is approached by a ramp about one foot (30 cm) long, formed by trampled grass, or through an access tunnel almost as long made by bending down surrounding straws and binding them with soft fibers. The rather similar nest of Hudson's Canastero often rests in a slight hollow beneath a cardoon bush. Walled and roofed with grass, it is lined with fine straws and rodents' fur, often with finely powdered horse dung as a bed for the eggs. The Sulphur-bearded Spinetail's ellipsoidal nest with a doorway in the side is built about a foot (20–40 cm) above the ground or water, amid sedges or other aquatic vegetation. Composed of grasses and rushes, it is softly lined with the sheaths of rushes, feathers, and down.

Two ovenbirds build their nests above standing water in ponds, ditches, or brooks. The Curved-billed Reedhaunter binds up to sixteen cattail stalks into the walls of its ellipsoidal nest, which may be nine inches high by six inches in diameter (23 by 15 cm), with a visor-

Curved-billed Reedhaunter *Limnornis curvirostris* Sexes similar Southern Brazil to central Argentina

like extension above the round doorway in the side. Vertical strips of cattail leaves cover the walls; cotton, cattail fibers, and where available, fragments of plastic form a bed for the eggs. This nest closely resembles that of the Marsh Wren of North America. Of similar form but more neatly finished is the nest of the Wrenlike Rushbird, in Chile called El Trabajador (The Worker) in recognition of its artisanlike diligence. This "wonderful structure" (the words are Hudson's, who knew it well) is firmly attached to three or more upright stems. The rushbird apparently daubs grass leaves with wet clay before ingeniously weaving them into its fabric, possibly with the addition of some kind of mucilage. Softly lined with feathers, the nest is light but strong and impervious to water; with a visor shielding the doorway in the side, it keeps its occupants snug and dry. Until the supporting rushes die and fall, the nest remains firmly attached to them, and in winter offers a comfortable refuge for small frogs, three or four of which sometimes crowd together.

Amid dense concealing vegetation from the ground up to about five feet (1.6 m), Des Mur's Wiretail builds a globular nest opening sideward. The walls are composed of dry grasses, herbs, twiglets, and sheep's wool; the interior is softly lined with feathers. When a parent enters to incubate or brood, it coils its long, wiry tail feathers inside instead of permitting the ends to project through the doorway, in the manner of an incubating male Resplendent Quetzal's longest plumes.

Several ovenbirds of different genera build nests of moss. In the mountains of southeastern Brazil, the Itatiaia Spinetail makes a domed nest with mossy walls several layers thick and a doorway facing obliquely upward. In an extremely wet region high in the mountains of Bolivia, François Vuilleumier found a Black-throated Thistletails' nest lying on top of a clump of grass beside a small shrub. Composed of sphagnum moss strengthened by twigs, the ovoid structure, eight by six inches (20 by 15 cm), had a doorway in the side but no entrance tunnel. It was lined all around with soaking wet moss, upon which lay two almost naked nestlings. Under a cold, penetrating rain, both parents brought food to them.

In rainy montane forests of Costa Rica, little Red-faced Spinetails attach their bulky nests to slender, drooping, leafy branches or vines, hanging free at the edge of a small opening in the forest or at its edge, sixteen to forty feet (5 to 12 m) above the ground. The globular structure, a foot (30 cm) or more in diameter, is composed largely of green

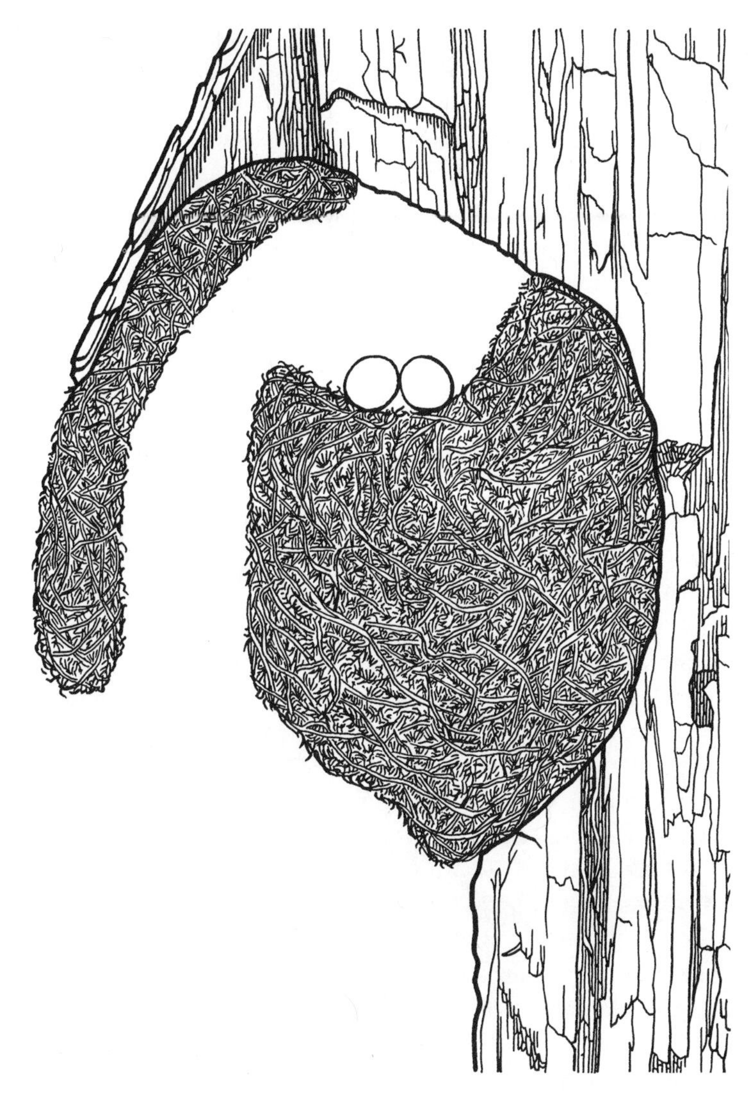

Nest of Spotted Barbtail *Premnoplex brunnescens* attached to side of mossy log bridging rivulet, sectioned vertically to show structure (from author's field sketch)

moss, bound together by slender vines, the ends of which often dangle loosely beneath it. The nest's attachment is strengthened by giving additional vines a turn or two around the supporting twig. From the inconspicuous orifice near the bottom, a passageway leads up to a chamber lined with dry leaves. When used for breeding, the nests are built by both parents. The nests in which a single spinetail sleeps may be made by it alone.

Attached to the side of a massive, decaying log that served as a precarious footbridge over a shallow brook, which flowed through a deep, wooded ravine in the highlands of Costa Rica, I found a nest that resembled no other that I have ever seen or read about. I could not imagine who built it until a Spotted Barbtail darted out as I passed by. The bulky structure, 8½ inches high and 7½ inches from front to back (22 and 19 cm), was composed almost wholly of green mosses and liverworts, bound together by fine, dark-colored rootlets. A flap of bark partly covered the front and helped to hold the nest in place, but this seemed inadequate to keep it in a shallow depression in the trunk. The nest blended well with the mosses, ferns, and other green epiphytes that covered the trunk around it. A vertical tube about 3¼ inches long by 1½ inches in diameter (8 by 4 cm) led upward from the nest's bottom to a chamber in the top, above the main mass of moss, where two white eggs rested in a shallow depression. I could see them only by inserting a flashlight bulb and a small mirror. The chamber's roof was the decaying wood of the trunk; when, after loss of the eggs, I removed the nest from its precarious situation, I found it penetrated from top to bottom by a hollow.

Unique among the multiform constructions of ovenbirds, and quite unlike the earthy urns made by several kinds of swallows, are the nests, miniatures of the baking ovens widespread in Latin America, fashioned of clay by the six species of horneros. Best known are those of the Rufous Hornero of southern and central South America, east of the Andes. Weighing up to about nine pounds (4 kg), the heavy, domed structure needs a firm support, such as the top of a fence post, the stout horizontal limb of a tree, the roof or broad beam of a building, rarely the ground. Occasionally a second, even a third, nest is built atop the first when this is in a favored site. By turning inward, on either the right or the left side, the wall at the front, where the oval entrance is situated, the builders make an antechamber or vestibule, separated by a partition with a doorway at the back from the inner chamber, where the eggs lie on a bed of dry grasses and feathers. Although a human hand readily enters the antechamber, it cannot be twisted around to reach the eggs in the inner sanctum. Like the clay of an adobe wall, that of the ovenbirds' nest is strengthened by a binder, such as horsehair, fibrous rootlets, straw, or dry cow dung. Often a builder carries a strand of hair or thread to a puddle and works it into a pellet of mud the size of a filbert before carrying it to the nest. The walls average about an inch (2.5 cm) in thickness. A typical oven measures nearly nine inches in height and slightly more in breadth at the base (22 by 23 cm), which is oval rather than perfectly round.

Although an ovenbirds' nest may stand intact for years, the birds build a new one each year, leaving their old nests to be occupied by other birds. Both sexes build, beginning in favorable autumns and continuing through the winter during spells of mild, wet weather when mud is available. Some nests are finished early in winter, others not until spring, depending upon the weather and the condition of the birds, as Hudson found. The horneros studied by Fraga started to build two or three months before they began to lay, but if they lost their first nest they could finish a replacement in about twelve days. Rarely, fledgling horneros help their building parents by carrying mud or dung to the nest, sometimes starting only nineteen days after they fledge. During six discontinuous hours of observation over an interval of six weeks, the adult male made eleven trips to a nest, the female five trips, one juvenile six trips, and the other four trips. Probably young horneros would help at more nests if their parents did not oppose their visits to unfinished structures, although the juveniles are permitted to remain nearby for months. Building with mud or wet clay is a skilled occupation, for until it has dried and hardened it is readily displaced by careless approaches, as happens when young White-winged Choughs in Australia help build bowls of clay. This may be the reason why horneros do not encourage their juveniles to participate in building.

The other five species of horneros have not received as much attention as has the familiar Rufous Hornero. The Pale-legged Hornero, widespread in northern South America, builds a typical clay oven or occupies an abandoned oven of the same form. In northeastern Brazil, Anita Studer and Jacques Vielliard found Wing-banded Horneros nesting inside epiphytic bromeliads or beneath the roof of a house. Their open cups were loosely made of dry grasses, cotton, and other vegetable materials, and lined with feathers, hair, and scraps of paper or plastic. This species is also reported to occupy holes in trees. Whether the Wing-banded Hornero's failure to build a clay oven is a primitive or a derived character is not known.

After our survey of nests of diverse forms, materials, and sites, we turn to those most characteristic of ovenbirds, the baskets, cabins, castles, and many-chambered mansions made of interlocked sticks and variously furnished. Few other birds anywhere construct their nests in this manner; in temperate North America, only the little Verdins of the southwestern deserts build enclosed nests of twigs somewhat like those of certain ovenbirds. Although the construction of all these nests is similar, their forms vary greatly. We can trace a series from nests that open upward, and appear to be the most primitive, through nests with a simple opening in the side, and those in which this is approached through a tunnel or tube, to structures of still more elaborate design. These diverse forms, now found in different genera, suggest the steps by which the most complex nests evolved.

Among the stick nests of simplest form is that of the Larklike Brushrunner, built in a shrub or small tree in northern Argentina, Paraguay,

Pale-legged Hornero *Furnarius leucopus* Sexes alike Northern South America to Bolivia and central Brazil

and southeastern Bolivia. Such a structure could be formed by continuing to build upward the walls of a typical cup-shaped nest, until they become a vertical tube leading inward from a skyward-facing opening. By thickening the wall slightly more on one side than on the other, the entrance tube is offset from the center of the cup, but not enough to prevent rain from falling directly into it, which may not be detrimental in the rather arid regions where these brushrunners live. The thick walls are composed of spiny or thornless twigs from two to five millimeters thick, mixed with which, near human habitations, are bits of wire, insulated copper cables, pieces of plastic cord, and the like. The chamber is lined all around with wool, feathers, reeds, thread, etc., on a base of pulverized dung. A typical nest was fourteen inches high by seventeen inches wide (35 by 42 cm). The vertical entrance tube was ten inches long and about four inches wide (25 and 10 cm). The chamber to which it led was 5½ inches (14 cm) in diameter, with walls about four inches (10 cm) thick. Another nest had a curving entrance tube about sixteen inches (40 cm) long.

A rather similar nest is built in a tree by the Short-billed Canastero, one of a number of species of *Asthenes* called canasteros (basket-makers) because of the forms of their nests. Slightly smaller than the brushrunner's nest, it is entered from above through a narrow, slightly curved or spiral tube. The chamber at the bottom is only 3½ inches (9 cm) wide. The entrance tube and chamber are lined with feathers of chickens, tinamous, and doves, or with rodent hair. In the chamber the lining is thickest and rests upon a layer of vegetable debris, earth, and lichens.

Best-known of the ovenbirds that build with sticks is the Firewood-gatherer of Argentina, Uruguay, Paraguay, and southern Brazil. Its nest, from twelve to forty-four inches (30 to 110 cm) long, is built low in a bush or up to sixty-five feet (20 m) high in a tree. These birds prefer a shrub or tree with few branches and leaves that might impede their access to the nest with twigs held transversely in their bills. If the site is attractive to them, it matters not how exposed to wind or near human habitations it might be, for the leñateros are almost fearless of people, and they will continue quietly to build beside a corral where shouting men are lassoing a herd of half-wild horses. The long axis of this somewhat bottle-shaped structure is inclined, so that the entrance faces obliquely rather than straight upward and rain cannot fall directly into the chamber. In the larger nests a crooked or spiral tube, up

Firewood-gatherer *Anumbius annumbi* Sexes alike Central Brazil to southern Argentina

to thirty-one inches (80 cm) long, leads from the doorway slantingly down to a chamber about seven inches (18 cm) in diameter. Occasionally, a nest has two entrances, as do rare nests of spinetails (*Synallaxis*). Composed of spiny or spineless twigs, tightly interlaced, the structure is lined with feathers, down, wool, rabbit skin, snakeskin, or inflorescences.

To carry large sticks to high nests is heavy work for the weak-winged, ambulatory, eight-inch (20 cm) leñateros. As they approach their nest with pieces balanced in their bills, many are knocked out by obstructing branches. The birds appear unable to fly straight upward with a load, and it seems never to occur to them that they might retrieve fallen sticks by carrying them far enough aside, then upward on a sloping course. Accordingly, they neglect fallen sticks while they go to a distance for others, with the result that a wheelbarrowful sometimes accumulates beneath a nest. Both sexes build, sometimes helped by offspring who may begin to bring contributions when only forty days old. After the breeding season, the big nest serves as a family dormitory.

Nests of the Lesser Canastero of southern Argentina and Chile are completely roofed over, leaving a round, laterally facing doorway in the side, near the top. Nearly spherical in shape, these small nests, built low in bushes, are six to eight inches (15 to 20 cm) high by slightly less in horizontal diameter. The shell of thorny twigs or softer materials is well lined all around with feathers, wool, or cotton. Thornbirds' nests, attached to drooping branches of trees, also have sideward-facing doorways, which often give access to an antechamber separated by a low sill or ridge from the inner chamber that contains the eggs. This construction gives to nests of Greater Thornbirds and Little Thornbirds a triangular shape when viewed from the side.

The next step in the elaboration of ovenbirds' stick nests is the construction of an entrance tube or tunnel as an extension from the side. Creamy-breasted Canasteros build amid the long, spine-bordered leaves of the great terrestrial bromeliad *Puya raimondii* on the high puna of Peru. Made of tightly interlaced twigs, their nests, twenty to twenty-four inches high and about sixteen inches in diameter (50 to 60 by 40 cm), consist of a spherical chamber reached by a tunnel about ten inches (25 cm) long and two or three inches (6 or 7 cm) wide, which rests horizontally on the puya leaves and opens outward from the plant. Tube and chamber are so well lined all around with feathers, wool, horsehair, and vegetable down that the interior remains perfectly dry under the hard, cold rains of these chilly heights.

Also equipped with entrance tunnels are the huge nests of cacholotes, apparently the largest built by solitary families of passerine birds. Those of the Brown Cacholote rest, horizontally or slightly inclined, at no great height in small trees with open branching. Composed of twigs up to twenty inches (50 cm) long and as thick as a man's little finger, these nests vary greatly in size. Some are over four feet (1.3 m) long and twenty-eight inches (70 cm) high. A nearly straight entrance tube, occasionally a yard or more (1 m) long, leads to a spacious vaulted chamber about one foot wide and half as high (26–35 by 14–20 cm). Unlike many of the foregoing nests, neither the entrance hall nor the walls of the chamber are lined. Only a thin bed of tiny, thornless twiglets covers the part of the wide floor where the eggs lie, often visible through the bottom of the nest. The similarly huge nest of the White-throated Cacholote is so strongly built that Hudson could stand on the dome of one and stamp upon it with heavy boots without injuring it in the least. With the roof removed, an eagle or a vulture, as he affirms, could nest quite comfortably in one of these structures.

Brown Cacholotes never breed or sleep in an old nest, which means that they frequently make new ones. In the Córdoba (Argentina) Zoological Garden, one pair followed by Ana and Manuel Nores built seventeen of their big nests in thirty-six months, at intervals that averaged sixty-two days. Although old nests are never reoccupied, they often serve as foundations for new ones, or sticks from them are transferred to the new constructions, which are then more quickly finished. Cacholotes take from fifteen to thirty-seven days to finish their nests, but like other ovenbirds they continue thereafter to add materials to them. In the absence of the parents, a juvenile may bring an occasional contribution to a nest, but like Rufous Horneros, the adults discourage this activity. Within their spacious chamber, Brown Cacholotes perform their nuptial rites in privacy, sheltered from predators—advantages not available to birds with open or small nests. This may, in fact, be the reason why Brown and the other two cacholotes build such big nests.

From the side of the Chotoy Spinetails' large ovoid nest of sticks, built in shrubs or trees, a straight or curved, horizontal or slightly inclined entrance tube projects abruptly for a foot or two (20 to 65, average 41 cm). About 4½ inches (12 cm) in diameter, with a passageway 1½ inches (4 cm) wide, it leads to a small chamber covered on the bottom with yellow vegetable floss and a few fresh leaves. Yellow-throated Spinetails build their nests about a yard above the water in aquatic plants, or at a similar height above the ground in nearby shrubs. At the side of the globular chamber the entrance tube curves so strongly upward that it opens skyward. Unlike spinetails and canasteros of colder climates, these birds of largely tropical distribution do not line their nests all around but only on the floor, where their eggs rest upon a mat of soft green leaves, tufts of wool, and a

Rufous Cacholote *Pseudoseisura cristata* Sexes similar Eastern Brazil, Paraguay, and Bolivia

yellow, cottony material. These two spinetails are closely related to *Synallaxis,* with twenty-six species the largest genus of ovenbirds, and builders of the most elaborate nests, most of which are entered through tubes or tunnels.

Since descriptions of the nests of all species of *Synallaxis* are not available, and to describe all of them would be prolix even if I could, I shall tell only about the three that I know best. The most widely distributed of these spinetails is the Pale-breasted, which ranges from central Argentina to southern Costa Rica. In different parts of this range, its nests have been found from 1½ to thirty feet (0.5 to 9 m) up in tangles or tussocks of coarse grasses, shrubs, and trees, but they are only exceptionally more than eight feet (2.5 m) above the ground. They consist of two parts: the nest proper, in a squat tower, and the entrance tube that projects horizontally from the rounded chamber in the lower part of the tower. The total length of these constructions varies from about twelve to twenty inches (30 to 50 cm), of which six to eight inches (15 to 20 cm) are occupied by the entrance tube. The towerlike part is twelve inches (30 cm) or less high and six or seven inches (15 or 18 cm) wide at the chamber, which is about four inches (10 cm) in diameter. When in bushy places, these nests are made largely of fine twigs, which the builders may break directly from

White-whiskered Spinetail *Synallaxis (Poecilurus) candei* Sexes similar
Northern Venezuela and Colombia

shrubs or gather from the ground; but in the grassy areas where Pale-breasted Spinetails often nest, woody twigs may be so hard to find that the birds use many straws and weed stems. Above the nest chamber an accumulation of coarser materials, including broad grass blades and straws, forms a loose, rain-shedding thatch. Fine, irregular fragments of downy leaves, plucked green, make a bed for the eggs; there is no other lining. A few scraps of shed snake or lizard skin, or fragments of cellophane, may be tucked here and there into the walls. A peculiar feature of several nests that I examined was a bristly collar, formed of small tufts of grass or other herbaceous plants with their short, stiff roots attached, around the mouth of the entrance tube. Possibly this discourages the entry of certain small animals.

Inhabitants of densely tangled thickets, such as those that in the humid tropics cover fields that have rested for a few years, Slaty Spinetails build more consistently with woody twigs, making more massive and slightly larger nests, of essentially the same form as those of Pale-breasted Spinetails. The entrance tube is usually straight and opens sideward, but it may be inclined sharply upward if supported by a branch that takes this direction. One nest with an upwardly directed

Nest of Pale-breasted Spinetail *Synallaxis albescens* in tussock of coarse grass

tunnel had roughly the form of a large coffeepot with the spout twisted to the left, as though by a fall. An exceptional nest had two entrance tubes, on opposite sides. I did not watch its construction, but I surmise that the two builders habitually brought materials from different directions, elongating the tunnels on the sides of their approach. This improved model, if so it may be called, did not contain eggs.

Slaty Spinetails' nests are situated from 1½ to fifteen feet (0.5 to 5 m) above the ground, in the low, dense thickets that these birds prefer, but often beside a path or small opening, or in a tree with dense foliage standing in a clear space near a thicket. Working together, a pair of spinetails seek on or near the ground smooth or thorny twigs from about two to six inches (5 to 15 cm) long, rarely over eight inches (20 cm). With immense labor, they maneuver thousands of these stiff pieces up through tangled branches, one at a time. When working fastest, a pair made about twenty-four trips to their nest in an hour. On reaching the nest, they pull and push the twigs about un-

Nest of Slaty Spinetail *Synallaxis brachyura* in orange tree. The ruler is one foot (30 cm) long.

til they fit firmly together in a coherent fabric that surrounds the brood chamber and, more thinly, the entrance tube.

This relatively enormous edifice is not, strictly speaking, a nest, but a cabin built to shelter a nest. The nest proper is a hemispherical cup, about 2½ inches (6 cm) in diameter, on the bottom of the chamber. Composed wholly of downy leaves, often of a species of *Solanum,* bitten and torn by the birds into fragments that are often jagged, and bound together with much cobweb, the cup can with care be lifted intact from the opened chamber of sticks. The female often starts to lay when little or none of the lining is present to cushion her eggs from the sticks. Without removing wilted leaves, the parents bring more and more of them while they incubate, until when the nestlings hatch they rest upon a thick, downy bed.

Above the sticks that form the chamber's ceiling, Slaty Spinetails pile broad, dry grass blades, fragments of bark or wood, and other coarse materials to form a thatch that may become seven inches (18 cm) thick—half the nest's total height—and protects the interior from the torrential rains that fall in the latter part of the nesting season in southern Central America. Shreds of reptile skin, stuffed here and there into the interstices between the sticks, add to the diversity of the nest's components. About three weeks may elapse from the beginning of building to the start of laying, but the nest is hardly finished at this time, for the lining of downy leaves is sparse or lacking. During the ensuing weeks the birds not only complete their complex household but devote much time to its maintenance. Nests built toward the end of the breeding season are smaller, with thinner walls and shorter entrance tunnels than those started earlier.

The most elaborate ovenbirds' nests of which descriptions are available are those of the Rufous-breasted Spinetail of northern Central America and southern Mexico. These nests have two unusual features: The entrance tube is surrounded by a platform of sticks on which the birds can walk; the entrance, instead of facing sideward at the tube's end, faces upward near the end of the platform. It is surrounded by a collar of twigs finer than those elsewhere in the edifice and often thorny. Sometimes the collar is high enough to suggest a low tower or chimney, down which the spinetail dives headfirst to reach the egg-chamber. This collar conceals the narrow orifice so well that I could not discover how the birds enter their nest until I watched

Rufous-breasted Spinetail *Synallaxis erythrothorax* Sexes similar About to enter its nest through the collar of fine twigs surrounding the doorway

one go in. The only other spinetail known to build a nest with a platform and upward-facing doorway is the White-whiskered, an inhabitant of arid country in northwestern Venezuela and adjacent areas of Colombia.

For its bulky castle of sticks the Rufous-breasted Spinetail needs a firm support, which it usually finds in a vine-draped bush or small tree in or beside the lush thicket where it dwells, often near a brook or wider river. A nearly horizontal limb with numerous lateral twigs serves well; two parallel branches, close together, are even better. Sometimes a network of interlacing vines offers an adequate foundation. Most nests are five to twelve feet (1.5 to 3.5 m) up; rarely one is built as much as twenty feet (6 m) high.

While building and attending their nest, members of a pair cooperate closely. To their chosen site they carry sticks to make a slight platform, to which they gradually add a flaring rim, until the structure becomes an open bowl, not unlike the still-unlined nest of many another bird, but much too wide for the spinetails, if they intended to occupy it at this stage. As the birds continue to add sticks, they deposit some of them a little short of the bowl's rim. Thereby they lengthen the rim, on the side from which they habitually approach along the horizontal supporting branch, into a projecting lip. Elsewhere, they build the walls upward and, finally, inward. As the birds add more and more sticks to the lip, it lengthens and broadens into a runway, along which they hop bearing contributions to their nest. As the walls of the bowl become higher, they are extended along the sides of the runway, converting it into a long, narrow trough. They begin to cover the structure at the point where the trough joins the bowl, and from here they extend the roof in both directions simultaneously, until the trough becomes a tube or tunnel, the bowl a roughly spherical chamber with a vaulted rather than a flat ceiling.

Some of the twigs that the spinetails find on or near the ground are twice or, rarely, thrice as long as the six-inch (15 cm) builders. The thicker twigs are about a quarter of an inch (6 mm) in diameter; the heavier ones weigh two or, rarely, 3½ grams. As a spinetail begins to fly up to its nest with a heavy stick, it is frequently dragged back to the ground by the weight of its load, but it persists until it gains a low branch in the tangle where the growing structure is situated. Then, by a series of hops and short flights from perch to perch, with frequent

Rufous-breasted Spinetails' nest viewed from above. At left, the thatch of coarse materials roofing the brood chamber; at right, the collar of fine twigs around the entrance

flitting of wings, skillfully maneuvering its clumsy burden through obstructing branches, it attains its goal and deposits its stick, then pushes and tugs until the piece fits snugly in the wall.

Often, however, an apparently unnoticed obstruction knocks a stick from the spinetail's slender bill, or it is dropped because it is too heavy. Then the tireless bird descends to retrieve the stick and undertakes anew the laborious task of raising it by a circuitous route to the nest. Likewise, the spinetail follows to the ground pieces that drop while it is arranging them. In this, Rufous-breasted Spinetails differ from Firewood-gatherers, who, as already told, permit fallen pieces to accumulate into a heap below their nests. Spinetails can ascend to their nests by a ladder of branches and vines; Firewood-gatherers must apparently fly with their burdens to their more exposed nests. This may account for the difference.

In somewhat less than two weeks a pair of Rufous-breasted Spinetails may completely cover their chamber and the approach to it, but their work is still far from done. The twigs above the chamber would not shed rain. To keep their eggs and nestlings dry, the birds work hard, bringing short, thick sticks, broad pieces of bark, weed stalks, dry petioles of the cecropia tree, fragments of broad leaf-bases of shellflowers, heliconias, and other giant herbs, and similar coarse pieces to form a thatch adequate to shed the hardest downpour. The loose materials are sometimes piled to a height of eight or ten inches (20 or 25 cm) above the chamber's ceiling.

A nest may appear completed after nearly a month's work, but the industrious builders, not yet satisfied with it, continue to add to it until the eggs appear, which in nests built early in the season may be five or six weeks after construction began. Even while they attend eggs or nestlings, the spinetails bring more materials to the edifice, increasing its bulk without altering its form, until it becomes a monument to the tireless industry of the little builders, disproportionate to their size. The nest's final dimensions depend, among other things, on how long it is occupied and the adequacy of its support. The bulkiest nests rest upon broad, firm foundations that retain everything the birds pile upon them. A narrow foundation may be loaded until additional sticks slip off, a weak one will sink down beneath the increasing weight of materials. Accordingly, nests vary considerably in size. A fairly typical structure was twenty-nine inches long and nineteen inches high (74 and 48 cm) at the end where the nest chamber was

Rufous-breasted Spinetails' nest with the top removed, showing the hallway carpeted with pale snakeskins leading from the entrance at right to the nest bowl, lined with downy green leaves, at left

situated. The chamber was five to six inches wide by 4¾ inches high (13 to 15 by 12 cm). The tunnel that led into the chamber was fourteen inches (36 cm) long and 1½ inches (4 cm) in internal diameter. The platform of sticks that surrounded it was eighteen inches long by seventeen inches wide (46 by 43 cm).

To furnish this wonderful example of avian architecture and industry, the builders cover the floor of their chamber with downy leaves, often of a shrubby *Solanum*. Unlike the fragmented, cobweb-bound leaves that line the Slaty Spinetails' nests, these remain whole and loose. Cast reptile skins, of which the birds seem never to have enough, are diligently sought, stuffed here and there into the walls, and laid on the entrance tunnel, often until they carpet the whole length of the birds' front hall. Most of these furnishings, as well as additional sticks and thatch, are brought after the eggs have been laid, even while the birds feed nestlings.

A pair of Rufous-breasted Spinetails who lost newly laid eggs promptly started another nest in a small acacia tree beside a river. This site proving inadequate, they began their third nest in a nearby willow tree. On the morning when the nest in the willow was at about the same stage as the unfinished nest in the acacia, I watched the birds at work. Both partners brought sticks to the willow-tree nest, but occasionally one of them would absentmindedly carry its contribution to the acacia. When this occurred, the bird in the willow called *pet pet pét chu* to guide its partner to the preferred site, and the latter took the next stick to the willow tree. More often, Rufous-breasted Spinetails work silently. This pair transferred many sticks from their first, pillaged nest to the replacement nest. I admired the way one partner guided the other.

All the nests already described are enclosed, either in structures built by the ovenbirds themselves, or in burrows, holes, crannies, or nests made by other birds. An exception is the nest of the Bay-capped Wren-Spinetail, which has attracted attention by its nonconformity to the rule in this family. The wren-spinetail usually breeds in the midst of a clump of rushes, sedges, or similar plants so dense that in effect it forms a roof above the nest, which is entered through a narrow gap between crowded stalks. Those who have examined these nests were uncertain how much building the occupants had done; the nests often appeared to be composed largely of detached fragments of the sheltering stalks lodged among their bases. Definitely brought by the birds were only the feathers and straws that composed the mat on which their eggs rested. Apparently, the Bay-capped Wren-Spinetail does not depart far from the family habit of nesting in enclosed spaces.

17 *Eggs & Incubation*

Since many ovenbirds are indeterminate builders, continuing to add to their nests after they have laid their eggs, there is no interval between the termination of building and the start of laying. Tropical species lay two eggs, less often three, rarely four. Sets of three and even four appear to be most frequent in spinetails of the genus *Synallaxis;* four eggs are occasional in nests of Pale-breasted and White-whiskered spinetails, and almost as numerous as sets of three in those of Rufous-breasted Spinetails. In the South Temperate Zone, clutches of three or four eggs are usual. Few ovenbirds' nests anywhere have been found with more than four eggs; the Chotoy Spinetail, which in Argentina lays from three to six, is exceptional. As in antbirds, flycatchers, manakins, and other nonoscine passerines, the eggs tend to be deposited on alternate days, but exceptions occur. Ovenbirds' eggs appear never to be speckled, blotched, streaked, or otherwise marked. Most are plain white, but some are tinted with green, greenish blue, or pale blue. The Wrenlike Rushbird's eggs vary from greenish blue to turquoise; those of the White-whiskered Spinetail from pale green to turquoise. Rufous-breasted Spinetails' eggs are white, faintly tinged with blue, or less frequently beautiful pale blue, those in the same clutch always alike, as far as I have seen.

As they have built their nest together, so the male and female share incubation. Although only a small fraction of the more than two hundred species of ovenbirds have been carefully watched at the nest, these few are sufficiently diverse to suggest that incubation by both sexes is widespread, if not universal, in the family. Watching nests where spinetails (*Synallaxis*) incubate is interesting and enjoyable.

These little birds are so confiding that one can sit comfortably, a few yards away and without concealment, while the varied activities at their big nests keep attention from wandering, as it is apt to do at nests of other birds where scarcely anything happens during incubation except widely spaced comings and goings of the incubating parent or pair. The greatest problem is to identify the sexes, which are alike. Sometimes a disarranged or missing feather helps to distinguish them, but usually at least one of them must be marked. By wrapping a tuft of cotton around the end of a twig, soaking it in paint, and sticking this improvised paintbrush into the twigs at the entrance of a nest, where the birds may touch it as they pass in or out, I have sometimes made spinetails mark themselves with identifying spots of white. However, this ruse may fail because the birds tug at the paintbrush until they pull it free and carry it away, without rubbing their plumage against it.

At a nest early in the dry season, situated in low, vine-draped trees on an islet in a mountain torrent, I watched a pair of Slaty Spinetails for ten hours. Spot, named for the white paint mark at the base of her bill, occupied the nest at night and was apparently the female. She took eight sessions in the nest, ranging from 18 to 44 minutes and averaging 31.1 minutes. The nine sessions of Tattertail, her partner with badly frayed tail feathers, varied from 4 to 44 minutes and averaged 27.7 minutes. The total time spent in the nest by each partner was exactly the same, 249 minutes; but since Spot's sessions averaged slightly longer than those of her mate, if I had timed an equal number of each bird's turns in the nest, she would have surpassed him. Moreover, she took the long night session. The eggs were unattended for five intervals, ranging from 2 to 27 minutes and totaling 50 minutes. The two parents together attended their eggs for 91 percent of the ten hours that I watched.

Often the spinetail arriving to take its turn at incubation hopped over the outside of the nest, sometimes making the circuit again and again, and crawling through a narrow space behind some twigs that projected from the wall of the chamber and touched the top of the tunnel. Then it entered the nest, and a few seconds later the partner who had been sitting would emerge. At other times the newcomer entered directly, without first inspecting the exterior of the nest. The changeover was usually made silently, more rarely with a rattling *churr.*

Like Rufous-breasted Spinetails, these birds continued throughout the period of incubation to devote much attention to the maintenance of their household. On most of their arrivals to incubate, they brought either a piece of downy green leaf for the lining, or a weft of cobweb to bind these leaves together after they had been torn into fragments. To fetch dry grass blades for the thatch and sticks for the walls, they made special trips; they rarely brought these items when they came to incubate. To see the contents of the chamber, I separated the sticks of a side wall to make a small hole. After my visit of inspection, I closed this aperture, adding fresh sticks when necessary. No matter how carefully I tried to obliterate all traces of my intrusion, I could never do it to the owners' satisfaction; their sharp eyes would always notice the spot where I had opened their nest. For days after the breach had been perfectly mended, they would continue to be preoccupied with it, stuffing tufts of spiderweb and scraps of snakeskin into the place where it had been made. Other birds that repair their nests are certain weavers, nuthatches, and swallows; most passerines cease to be concerned about their structures after they begin to incubate, even when a little timely mending might save them from falling. Not only did these Slaty Spinetails take many things into their nest; occasionally, at the end of a session, one emerged with a stick or shred of withered leaf in its bill, and deposited this on the outside before it flew away.

This pair of Slaty Spinetails did most of their routine inspection and repair work late in the clear afternoons of the dry season. In five morning hours they neglected their eggs for only two minutes. In five hours of the afternoon they left their eggs alone for four intervals that totaled forty-eight minutes, during most of which time both parents attended to the nest itself, pulling up and tucking into place sticks that were slipping down from the walls and tunnel and blades of grass that were falling from the roof. They brought new materials for both roof and walls. On two evenings, the spinetails engaged most actively in tidying and repairing their nest after sunset, when the Gray-capped and Vermilion-crowned flycatchers who roosted on their islet had already gone to rest in the treetops above them. The spinetails continued to be very busy with their housekeeping, going over and over the nest, pushing in a twig here and there, now and again flying off to bring more pieces, until I could hardly follow their movements in the failing light. At last, when it was nearly dark, Spot retired into the chamber with the eggs, while Tattertail flew away, probably to roost in a bush or tree.

For this pair of spinetails, housekeeping included ridding their nest of insects. One afternoon Spot spent ten minutes rapidly picking off with her slender bill small brown ants that swarmed over the exterior of her nest and up and down the supporting branch. Whether she swallowed or dropped them, I could not tell. After the eggs hatched, I watched Tattertail engage in the same occupation. The sideward twitches of his head, each time he seized an ant in his bill, left no doubt that he tossed it aside instead of eating it. In the tropics, as well as elsewhere, ants devour many nestlings.

Many years later, I watched a Slaty Spinetails' nest, more favorably situated for observation than most, in a small, vine-draped tree growing in a fairly clear space amid a dense second-growth thicket. In 17½ hours, covering all parts of the day, I timed twenty-eight full sessions by both sexes, which ranged from 2 to 120 minutes in length and averaged 25.2 minutes. The eggs were unattended for twelve intervals of 1 to 48 minutes, averaging 13.0 minutes. Only one of the intervals of neglect exceeded 22 minutes; most were much shorter. During the active period when the birds were taking turns in the nest, they were inside with the eggs for 81.9 percent of the time.

Although from my paintbrush the female had acquired a white spot on her slaty breast, I sometimes failed to see this in the dim light of a heavily overcast afternoon. For comparing the participation of the sexes in incubation, only twelve hours of my record are available. In these hours, the male took twelve sessions, ranging from 2 to 120 minutes and averaging 33.8 minutes. The female's nine sessions varied from 9 to 42 minutes and averaged 22.6 minutes; but to compensate for her smaller share in incubation during the morning and early afternoon, she began her nocturnal session at 4:11 P.M. on the rainy afternoon, at least two hours before many of the surrounding birds became inactive in the evening. This session lasted until 5:34 next morning, or for 13 hours and 23 minutes.

In fifteen hours of active nest attendance, this pair brought green leaves for the lining sixteen times, coarse pieces for their thatch four times, snakeskin twice, one stick, and unidentified items six times, a total of twenty-nine trips. This does not include the fallen materials that they retrieved from the tangle of vines beneath the nest and carried back to it. Moreover, they gave much attention to the neatness of their mansion. This housekeeping was usually done by the bird who had just been relieved from a spell of incubation by the arrival of its partner, but sometimes the incubating bird would emerge without waiting for relief and turn its attention to the nest. Rarely both partners were so occupied simultaneously. The bird in the nest rattled loudly when it heard its mate, foraging or seeking materials beyond view.

At two nests of the Rufous-breasted Spinetails more briefly watched, one in Honduras and the other in Guatemala, the sexes did not share incubation as equally as the Slaty Spinetails did. In approximately five hours at the first of these nests, the marked partner who occupied the nest at night took six sessions ranging from 6 to 40 minutes and averaging 24.8 minutes. Her mate sat only thrice, for 14, 14, and 18 minutes. The eggs were alone for five intervals of 3 to 36 minutes, averaging 18.6 minutes. Excluding sessions and absences that I did not watch in full, the female was with the eggs for 149 minutes, the male for 46 minutes, and they were unattended for 93 minutes. The eggs were incubated for 67.7 percent of the time. At the second nest, also, one partner incubated about three times as long as the other, but they covered their eggs more constantly, for 81.5 percent of four hours. As at nests of the Slaty Spinetail, the sitting partner often remained in the nest until its mate had entered, but sometimes, on hearing the other approach, it emerged and the two met outside. At times, the bird in the nest called loudly in response to the notes of its mate beyond view in the vine tangle.

In five hours at the first nest, the female brought green leaves nine times and one stick. Her partner, who incubated less, brought much more material: sticks, five times; green leaves, twice; pieces for the roof, twice; snakeskin, seven times—sixteen total contributions. This bird evidently found a treasure trove of snakeskin, for in midmorning he came to the nest four times in quick succession, each time bringing a large, limp shred. Rufous-breasted Spinetails seem never to have enough of these shed skins. If the bird arriving for a turn of incubation does not bring a stick or a green leaf, it will probably have a fragment of reptile skin in its bill, to be pushed into crevices on the outside or carried inside. Some is taken into the nest chamber, but much more is deposited on the floor of the entrance tunnel, sometimes enough to cover its whole length. Moreover, the spinetails do not permit these precious exuviae to rest long in one place. Frequently they shift them from one spot to another on the nest's exterior; they carry a piece from the outside to the inside; or they bring out a fragment from the interior. When it ends a session of incubation, a spinetail often climbs through the entrance chimney with a shred of skin that it will deposit somewhere on the outside of the nest before it flies away to forage.

Many other birds, including other spinetails and, especially, hole-nesters such as Great Crested Flycatchers and house-wrens, place snakeskin in their nests, but no other bird that I know uses it as lavishly as do Rufous-breasted Spinetails. With them, seeking these pale exuviae and shifting them around has become a pastime or an obsession that, apparently, neither adds to the comfort or stability of their nests nor increases their success in reproduction. Although the partner who is off duty is chiefly concerned with care of the nest and the search for reptile skin, frequently the pair leave their eggs unwarmed for many minutes while both engage in this absorbing occupation.

In one Rufous-breasted Spinetails' nest was a nook or niche, just large enough for the birds to turn around in it, between the thatch, the front wall of the chamber, and the top of the tunnel. This cranny attracted the birds irresistibly; on arriving or before leaving the nest, and often on their rounds of inspection, they disappeared in it for a few seconds. Here they deposited, briefly or for a longer time, some of the leaves or shreds of reptile skin that they brought to their nest. I could not discover that this niche had any special significance in the nest's economy; it appeared to be an accident of construction, for in other nests of this species I failed to find a similar nook. Apparently, the birds' behavior was merely a manifestation of their habit of pushing into crannies. Spinetails are as strongly attached to the nest itself as other birds are to the nest's occupants. The time and energy they lavish on it is impressive. The search for food leaves them ample time for housekeeping, to which they attend with zeal.

For twelve hours, including all of one morning and the whole afternoon of another day, I watched a nest of Pale-breasted Spinetails whose sexes I did not distinguish. As in the preceding species, both incubated. Usually the bird arriving for a turn on the eggs flew silently into the nest before the other left. Ten completed sessions of both ranged from 29 to 102 minutes, totaled 592 minutes, and averaged

59 minutes. The eggs were unattended for only three intervals, of 27, 18, and 29 minutes—total 74 minutes. One or the other member of the pair was in the nest for 90 percent of the eleven hours comprising the sessions and absences that I timed in full.

After these spinetails began to incubate steadily, they gave much less time to tidying their nest, bringing new materials, or shifting those already present than did the two spinetails with larger or more elaborate nests. Rarely the new arrival would spend a minute or less going over the outside of the nest and putting it in order before it entered, and the partner just relieved would devote an equally brief interval to the same task before it flew away. During the twelve hours, they brought only three pieces to the nest: a green leaf and two others that I could not identify. One of them carried a piece of snakeskin from the nest when it ended its spell of incubation. After flying to a distance, the departing bird might deliver a long rattle, or sometimes a sharp *bip bip bip bip*. In contrast to their sparse utterances after their set of two eggs was complete, while in the nest with their first egg they were surprisingly voluble. Part of the series of 1,300 *bet chus* mentioned in chapter 15 came from inside the nest. This nest was above a low bank beside a muddy, unpaved road traversed by jeeps and heavy trucks with loudly clanking tire chains, but the spinetails sat steadily through all the racket so close beside them.

Learning how Buff-throated Automoluses incubated in deep burrows was less entertaining and far more exacting than watching spinetails at their castles of sticks. If I did not keep my eyes steadily fixed for long hours on a round hole in a bank, I might miss a brown bird's swift inward or outward dart, and so break the continuity of my record. At my first nest, beside a woodland stream, I watched for twelve hours without seeing a single changeover; but after the nestlings hatched both parents brooded them, providing a strong presumption that both also incubated, for the individual bird who engages in one of these activities nearly always participates in the other. At this nest, four sessions on the three eggs, two not timed in full, lasted 78, 183+, 124+, and 72 minutes. Two intervals of neglect were 64 and 52 minutes. Thus, this pair attended their eggs for 81 percent of their active period. These birds approached their burrow through the light woods behind the bank; on leaving they flew downstream toward neighboring heavy forest.

In eighteen hours of watching at a later nest, I only twice saw one automolus come to replace the other. The record that I made during the first twelve hours is short enough to be given in full.

May 6, 12:30 P.M. I enter the blind; bright sun, clouding over. Shower falls at 2 P.M.

2:48. *An automolus suddenly leaves the burrow, flying out across the edge of the pasture in front of it.*

4:11. *An automolus arrives through the thicket behind the burrow and silently enters.*

6:05. *I leave the bird in the burrow in the failing light. Rain fell hard during the late afternoon.*

May 7, 5:10 A.M. I resume my watch at dawn.

5:21. *An automolus leaves the burrow, flies silently over edge of pasture.*

5:53. *An automolus enters, voicing only a few low notes.*

6:55. *The mate arrives with bill full of material (rachises of mimosa?), clings to vertical stem of sapling in front of the burrow. The one that has been incubating darts out and away. The new arrival is alarmed when I too suddenly raise my binocular to the window of the blind. It retreats into the bushes behind the burrow and stays there for 25 minutes, moving around mostly out of sight and constantly voicing rattling notes.*

7:20. *At length this bird enters the burrow.*

8:23. *It darts out and away.*

9:37. *A bird silently enters the burrow.*

11:35. *It darts out and away. I leave.*

On a morning a week later, these birds incubated no more constantly. In eighteen hours at this burrow, six diurnal sessions on the eggs ranged from 62 to 138+ minutes (the longest was begun before I arrived to watch) and averaged 96.8+ minutes. Five intervals of neglect ranged from 32 to 122 minutes and averaged 69.2 minutes. The eggs were incubated for only 58 percent of the eighteen hours. Once,

Buff-throated Automolus *Automolus ochrolaemus* Sexes similar Southern Mexico to northern Bolivia and Amazonian Brazil

after the nest had remained unattended for two hours, I was in front of the burrow, looking in with a flashlight to assure myself that neither parent had darted in unnoticed by me, when one of them shot out of the bushes behind the burrow and almost bumped into me. On some of my later visits of inspection, I found the attendant parent just inside the burrow's mouth, to which it had apparently advanced on hearing or feeling my approaching footsteps. When the flashlight's beam fell on its face, it retreated to the inner end where the eggs lay. Then no moderate amount of stamping on the ground a few yards away would send it into the open, and I could not see whether the eggs had hatched.

After a pair of Scaly-throated Leaftossers had incubated for about a week in the wall of clay heaved up on the roots of a fallen tree, I passed a morning watching them from a blind. Twice, after one had been sitting for about an hour, its mate entered the burrow, and then one flew out. Since I could not distinguish the sexes, I could not exclude the possibility that the new arrival had departed after a brief visit to its incubating partner, but this was much less probable than that it had replaced the other, far back in the burrow where I could not see them. Accordingly, I concluded that both sexes incubate, as in other ovenbirds. Also like other ovenbirds, they do not keep their eggs continuously covered. The four sessions that I timed in full lasted 71, 54, 53, and 59 minutes. The eggs were neglected for intervals of 34 and 38 minutes. They were incubated for 76.7 percent of the morning. On one occasion, the leaftosser arriving to begin its turn on the nest brought a big sheaf of material.

Often, as I walked down the seldom-used cartroad toward the nest, a leaftosser, hearing my footsteps, or feeling vibrations, came to the entrance of the burrow and looked out. Here it stood until I approached to within a few yards, then darted forth and flew down the roadway ahead of me, low above the ground, for a good distance before it turned aside into the forest undergrowth and vanished. At other times it would alight on the road in front of me, then as I approached fly onward a few yards to alight again, repeating this several times before it veered into the undergrowth.

It did not take me long to learn that both sexes of the Plain Xenops incubated in the low hole carved by Olivaceous Piculets in a forest clearing. I began my vigil at dawn, when I could hardly distinguish the tiny round orifice in the side of the stub. As daylight increased, a xenops called at the edge of the woods, and at the same moment a white-striped face appeared in the doorway in front of me. The owner of this face, doubtless the female, flew from the nest into the forest, and then her partner entered to incubate. Here he remained quietly until, seventy-two minutes later, the female flew from the woods to the doorway and clung beside it. Thereupon, the male climbed out of the cavity and flew toward the forest, rapidly repeating a sharp *chip* as he went. His spouse promptly entered the hole.

At the forest's edge, the male clung to a dead twig and sang his clear, sharp, rapid trill over and over. After a while, his mate began to answer, trilling in the nest. After she had been inside for only ten minutes, he returned to the doorway and trilled, while she responded with a more subdued trill from within. After two minutes of intermittent

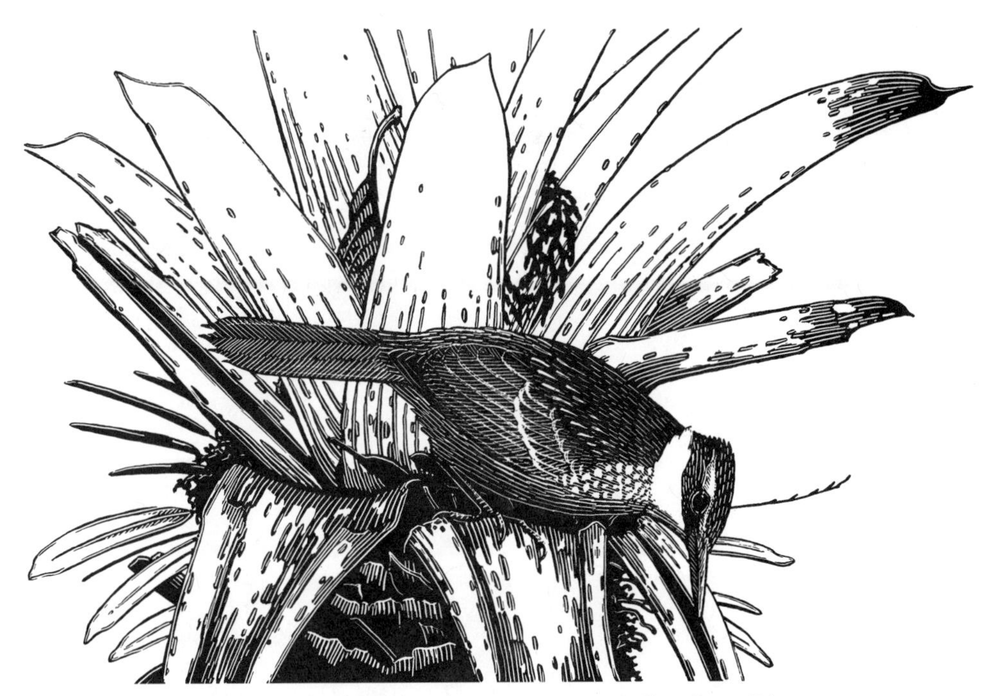

Buffy Tuftedcheek *Pseudocolaptes lawrencii* Sexes similar Costa Rica to Ecuador

duetting, the male replaced her in the cavity. Half an hour later, she came silently to resume incubation, and he flew, chipping, back to the woods. She sat until, three quarters of an hour later, he alighted beside the doorway and called her out with soft trills. Voicing low notes, she vanished among the trees. Thus, during the first half of the forenoon, the two partners, sitting alternately, attended their eggs continuously.

After midmorning, the attendance of this pair of xenops changed abruptly. Through the remainder of the day, long intervals of neglect alternated with long sessions on the eggs, and I never again saw one member of the pair replace the other. In eleven hours of watching, I timed nine completed sessions of incubation, ranging from 12 to 118 minutes and averaging 49.8 minutes. Four intervals of neglect varied from 25 to 51 minutes and averaged 43.5 minutes. The eggs were attended for 72 percent of the daytime. Years later, at a second-brood nest twenty feet (6 m) up in a tree, I found the same inconstant incubation. To leave their nest cavity, these xenops dropped almost straight downward to near the ground before they leveled their course and flew off through the underwood. This mode of departure, widespread among small birds that nest high, may make the point of origin

TABLE 6. *Incubation and Nestling Periods of Ovenbirds*

Species	Incubation period (days)	Nestling period (days)	Locality
Blackish Cinclodes	ca. 16	ca. 25	Falkland Islands
Brown Cacholote	18–20	18–23	Argentina
Rufous Hornero	16–17	24–26	Argentina
Pale-legged Hornero	16–17	26 (33)	Peru
Wrenlike Rushbird	16	15	Argentina
Tufted Tit-Spinetail	14	15–16	Argentina
Pale-breasted Spinetail	15–16	—	Argentina
Slaty Spinetail	18–19	17 (+?)	Costa Rica
Plain-crowned Spinetail	18	—	Suriname
Rufous-breasted Spinetail	17–18	—	Guatemala
Rufous-fronted Thornbird	16–17	21–22	Venezuela
Freckled-breasted Thornbird	16	—	Argentina
Firewood-gatherer	16–17	21–22	Argentina
Buffy Tuftedcheek	ca. 24	ca. 29	Costa Rica
Buff-throated Automolus	20–22	(16)–18	Costa Rica
Plain Xenops	15–17	13–14	Costa Rica
Scaly-throated Leaftosser	21 (+?)	15	Costa Rica

Note: Less usual periods are given in parentheses.

of their flight less obvious to predators. At this nest, as at the earlier one, the bird coming for a spell of incubation sometimes brought a billful of fine fibers to tuck beneath the eggs.

We have many descriptions of ovenbirds' nests, often with excellent illustrations, especially from Argentina, where the family is well represented. Nevertheless, few people have sat long to watch what happens at these nests and reported their observations. Since Hudson's pioneering studies in Argentina over a century ago, if not earlier, it has been known that both sexes of certain ovenbirds incubate, but the only available details, other than my own, we owe to Rosendo Fraga, who studied the Rufous Hornero, and to Ana and Manuel Nores, who watched the Brown Cacholote. In fourteen hours of watching, distributed between five nests of Rufous Horneros, these banded pairs incubated, or at least remained out of sight in their ovens, 72 percent of the time. The longest session that Fraga recorded was 158 minutes; the longest interval of neglect, 31 minutes. By day, the sexes of the cacholotes studied by the Nores incubated with equal constancy, taking sessions that ranged from 5 to 35 minutes and averaged 28 minutes, and never neglecting their eggs for more than 30 minutes at a stretch. On most nights, the female alone attended her eggs, while her mate slept nearby in the best preserved empty nest or amid foliage.

The only ovenbird who, as far as I could learn, incubated alone was a Buffy Tuftedcheek, who occupied her inaccessible hole nearly three quarters of a morning, or about as much as pairs of other species incubating alternately do. The most continuous incubation known to me was by Rufous-fronted Thornbirds, as will be told in chapter 19. The known incubation periods of ovenbirds range from about fourteen to twenty-two days (table 6).

18 The Young & Their Care

Ovenbirds hatch with tightly closed eyes, like other passerine nestlings. The pink skin of spinetails, xenops, leaftossers, and automoluses is sparsely shaded on the upperparts with grayish down, but that of horneros is quite naked. The interior of the mouth is yellow in spinetails, horneros, thistletails, and cinclodes, but flesh-color in burrow-nesting leaftossers and automoluses. The mouths of leaftossers, like those of many other passerines, are bordered by yellow flanges that make their gapes more conspicuous, but flaps of skin are lacking on automolus nestlings. The parents promptly remove the shells from which the nestlings have escaped.

Both parents feed the nestlings, mainly with insects and their larvae, spiders, and other small invertebrates. Crickets, mole crickets, larvae of soil beetles, and earthworms enter largely into the diet of nestling Rufous Horneros, whose parents hunt their food on the ground. The Buff-throated Automolus adds a few small lizards to its nestlings' fare; the Streaked-breasted Treehunter gives its young many lizards and a few little frogs as well as insects and spiders. When a Rufous-breasted Spinetail offered a small berry to its fledgling, the youngster rejected it and the parent swallowed it. On no other occasion did I see an ovenbird of any species bring fruit to its offspring, but the burrow-nesters flew so swiftly into their tunnels that it was usually impossible to identify what they carried. As far as I could see, parent ovenbirds of several species brought only one item at a time, held in the tip of the bill.

Meals of nestling ovenbirds tend to be infrequent but substantial. In watches of three or more often four hours (total thirty hours) at nests of Scaly-throated Leaftossers, Buff-throated Automoluses, Streaked-breasted Treehunters, Plain Xenops, and Slaty Spinetails with nestlings of various ages, I recorded feeding rates of 1.0 to 2.5 meals per nestling per hour. Rufous-fronted Thornbirds fed their young more frequently, up to 6.7 times per nestling per hour. Fraga's Rufous Horneros fed day-old nestlings at the rate of 2.3 times each per hour, sixteen-day-olds at the rate of 8.8 times per nestling per hour. Dyrcz's Pale-legged Horneros were fed at rates from 0.85 times per nestling per hour when they were a day or two old to 2.85 times after they were older. In the Falkland Islands, where food was abundant, two ten-day-old Blackish Cinclodes were fed 30 times in an hour, or 15 times for each, but such a high rate is exceptional even in this species. Even while feeding and brooding nestlings, spinetails with big nests of sticks find time to bring more material and keep them tidy.

As they have shared building, incubation, and feeding their young, so both parents take turns brooding them. Except probably at high altitudes and high southern latitudes, for which observations are lacking, young ovenbirds appear to need little brooding in their covered nests. Diurnal brooding of Rufous Horneros continued, in decreasing amounts, until they were twelve days old, but in their secure ovens a parent accompanied the nestlings at night until they were fifteen to seventeen days old and well feathered. Buff-throated Automolus nestlings are brooded by day until they are ten days or less of age. I have not known nocturnal brooding to continue longer than this, and some week-old nestlings slept alone. Such early cessation of nocturnal brooding is widespread in burrow-nesters, including the automolus's neighbors in a bank, the Blue-diademed Motmot and the Southern Rough-winged Swallow. Deep in the warm earth, nestlings still nearly naked can retain enough heat without the parental coverlet; and by sleeping elsewhere, the parent avoids the risk of being trapped in the burrow by a marauder that blocks the entrance—a disaster that would prevent the replacement of the lost brood.

Ovenbirds keep their nests clean by swallowing the nestlings' fecal sacs or carrying them away in their bills. Each dawn, when I looked into a Buff-throated Automolus's burrow with a flashlight, I saw a row of five or six pure white, round, little pellets lined up on the front rim of the nest bowl. During the early morning the parents carried away these droppings, one each time they visited the nest with food, until all this waste had been removed. Scaly-throated Leaftossers carried away the fecal sacs of younger nestlings, but after they were older and often came to the burrow's mouth to receive their meals, making it unnecessary for the adults to enter, they neglected sanitation, with the result that the inner end of the burrow became foul and swarmed with maggots. Similar relaxation of nest sanitation is not rare in hole-nesting birds.

Rufous Horneros' eyes begin to open at four or five days of age, but until they are nine days old they mostly keep their eyes closed. Scaly-throated Leaftossers also have open eyes at nine days. Plain Xenops are fairly well clothed with plumage when only nine days old, but larger ovenbirds become feathered more slowly: leaftossers at about twelve days; automoluses at two weeks; horneros at about sixteen days.

Nestling ovenbirds are voluble. Hudson noticed how garrulous young Rufous Horneros, when only half feathered, practice trills and duets in their secure abode, in shrill, tremulous voices, which change to the usual hunger cries of young birds when a parent enters with food. Half-grown nestling Scaly-throated Leaftossers raised a loud chorus of softly chiming notes each time they were fed. While the parents were still out of sight, the sound of their approaching wings would prompt these pleasant notes, which continued until the young had received their meal and the parents flew off, then gradually faded away. The rustle of wings of birds of other kinds flying past the burrow would release a similar outburst of calling; apparently, the nestlings could not distinguish the wing sounds of other birds from those of their parents. Occasionally a parent arrived with food while I was looking into the burrow. Even with a lighted bulb beside them, the sound of the parents' wings set off the nestlings' chiming clamor; they could not see the parent because my head screened the doorway. The surprised adult would drop down into the roadway near me and utter sharp notes of alarm.

The voice of an approaching parent Buff-throated Automolus would elicit a clear little chiming or trilling of united nestling voices that continued after the adult entered the burrow to feed them. By directing a beam of light into the tunnel and making a clicking noise with my tongue, I could provoke a similar outcry; but after the nestlings were two weeks old they would shrink back into the nest instead of calling when I peered in at them. A loud, resonant chiming greeted the approach of a parent Striped-breasted Treehunter with food and assured it that its progeny were still safe in their burrow. While resting in their nest cavity in a tree, Plain Xenops nestlings rhythmically repeated a low, sharp *peep*. High-pitched, sharp notes from nestlings unseen in a globular, mossy nest hanging high greeted the arrival of a parent Red-faced Spinetail with food. Unlike all these other nestlings, a Buffy Tuftedcheek in a hollow in a decaying trunk never uttered a sound audible to me. Apparently, the voices of the foregoing nestlings do not often attract predators to them; but the volubility of young Hudson's Canasteros is disastrous to them, as will be related in chapter 20.

Older nestling Rufous-breasted Spinetails call for food with loud, trilling chirps that penetrate the walls of their nursery. One afternoon, to examine the nestlings, I made a small gap in the nest's side and, contrary to my usual practice, purposely left it open to see what the parents would do. The moment I withdrew a few steps, a parent, who had been fidgeting around a short way off, flew up with a stick that it thrust into the breach. Then it continued with great industry to bring twig after twig to the gap, flying down again and again to fetch up sticks that had fallen beneath the nest, instead of seeking them at a distance. Meanwhile, the other parent brought an insect to the hungry nestlings. The discipline of these spinetails was admirable; neither of them was the least flustered nor uttered a sound. Without consultation, each knew exactly what the unusual situation demanded and did its part while I stood nearby, intently watching. When dusk was deepening into night, one parent was still working sticks into the nearly obliterated breach, while its partner busily arranged pieces on the platform. Like Slaty Spinetails, Rufous-breasted Spinetails continued for days to worry about the point where their castle wall had been breached. In addition to twigs, they stuffed into it green leaves and shreds of reptile skin, both of which they usually deposited elsewhere.

Although spinetails fearlessly repaired a small injury to their nest while I stood nearby, I have never known them, nor any other ovenbird, to try to lure me away by "feigning injury," to threaten me with feints of attack, or even to complain or protest vehemently when I visited one of their nests. While I held a nestling Rufous-breasted Spinetail in my hand, the watching parents made no outcry or demonstration, but as soon as I released it, one of them brought it food. Instead of fussing or complaining about a threat or injury, spinetails promptly set about to remedy the situation.

To learn how long undisturbed nestling spinetails remain in nests with long entrance tubes was difficult, for they would flee prematurely if I made an opening in the wall as the only way to see how many were still within. To prevent this, I blocked their exit with a handkerchief stuffed into the end of the tunnel while opening the chamber. I could dimly see the nestlings retreat down the tube. After I had examined the interior and closed the gap that I had made in the wall, I tried to make them return to the chamber by prodding them gently with a twig thrust through the wall. However, as soon as I removed the handkerchief, a sixteen-day-old Slaty Spinetail darted out and flew about twenty feet (6 m) without losing altitude, to alight in a bush. When I approached, it fled farther into the thicket. Its nestmate remained looking out from the end of the tunnel until a parent arrived with a large insect, whereupon it hopped out and fluttered to the ground. Without my interference, these fledglings would almost certainly have stayed in the nest until at least the following morning, for a nestling period of no less than seventeen days. Even when I could catch and replace in its nest a spinetail that had departed prematurely when I examined its nest, it refused to remain. Other ovenbirds have spontaneously abandoned their nests at ages of about fourteen to twenty-nine days (table 6).

A fifteen-day-old Scaly-throated Leaftosser that had just left its natal burrow flew fairly well but lacked control; every few feet it struck a vine or other obstruction in the forest undergrowth and was knocked to the ground. On its first flight, a Buffy Tuftedcheek covered about fifty feet (15 m) on a slightly descending course and promptly vanished amid vines densely draping a tree. After their first day in the open, young Rufous-fronted Thornbirds are led back to sleep with their parents in the nest, and the same is probably true of other ovenbirds, but observations are lacking. Neither parent nor fledgling Slaty and Rufous-breasted spinetails, Rufous Horneros, or Buff-throated Automoluses sleep in the nests from which the young have just flown. When about forty-five days old Rufous Horneros have become fairly competent foragers, almost independent of their parents. Forty-six-day-old Slaty Spinetails still received an occasional insect from their elders. Young horneros, Firewood-gatherers, and Rufous-fronted Thornbirds help their parents to build (chapter 16).

Some ovenbirds have long breeding seasons, especially at low latitudes and altitudes. In Suriname, Yellow-throated Spinetails lay eggs throughout the year; Plain-crowned Spinetails, from December to September; and Plain-breasted Spinetails, from January to October. In southern Costa Rica at 2,500 feet (760 m) above sea level, Slaty Spinetails lay from January to September, from the beginning of the dry season until well into the long rainy season. In these long breeding seasons, ovenbirds have time to rear several broods, but so many are lost that much of their effort is directed toward the production of a single brood of fledglings.

Although probably frequent, true second broods (as distinct from replacement nests) have not often been reported. In northern Venezuela, a pair of Rufous-fronted Thornbirds, accompanied by juveniles of their first brood, laid another set of eggs in the same nest, about thirty-five days after their first brood fledged. Also in Venezuela, a pair of White-whiskered Spinetails who had reared three fledglings had a clutch of four eggs in the same nest about twelve days later. Beyond the tropics in Argentina, Rufous Horneros not infrequently attempted second broods, laying again from six to nineteen days (mean 10.5 days) after their first brood flew. They also raised their second broods in the same nest as the first. In Chile, Common Miners and Dusky-tailed Canasteros attempt second broods.

Although ovenbirds may lay two or more clutches in the same nest in a single breeding season, they appear rarely to use the same nest in successive seasons. In the humid tropics, laboriously constructed nests of vegetable materials deteriorate so much during the rainy months that the birds must build anew when they resume breeding the following season. Even Rufous Horneros, whose massive ovens of clay often remain over winter sound enough to be occupied the following spring by birds of other kinds, do not themselves use them for more than one season. I have known Scaly-throated Leaftossers to nest four times in the same burrow, but for each brood they built a new nest in it. From May until December of the same year, they laid three sets of eggs, at least one of them after a successful earlier nesting. In the valley of El General in southern Costa Rica, these leaftossers breed into the very rainy last quarter of the year, when scarcely any other birds nest.

After leaving their nests when from eighteen to twenty-three days old, Brown Cacholotes were fed generously by their elders during the following twenty days, after which parental feeding was gradually reduced. Nevertheless, the young birds continued to beg for thirty-six to forty days after they left the nest. They remained in the parental territory for five to thirteen months, helping to guard the nests and adding sticks to them, as already told. Because parents were more tolerant of their daughters than of their sons, the latter departed at an earlier age. After their departure, 40 percent of the young birds claimed areas near their parents' territories, mated, and built their first nests. Of 105 cacholotes fledged during the Nores's four-year study, 42 (40 percent) reared fledglings, 9 when one year old, 22 when two years old, and 11 not until their third year.

19 *The Rufous-fronted Thornbird*

I have searched for years without finding a nest of a bird that I desired greatly to study, but as soon as I reached Venezuela to learn about Rufous-fronted Thornbirds, I knew that I would not have this difficulty. Their large, oblong nests of interlaced sticks hung in full view on sparsely foliaged trees and even on service poles along busy highways. They were much more conspicuous than the little, wrenlike birds that make them. About seven inches (17.8 cm) in length, these thornbirds are brownish olive above, tinged with rufous on the wings and tail. The chin and throat are whitish, behind which they become grayer. Lacking the cinnamon-rufous forecrown of southern races of their species, the northernmost form has been called the Plain-fronted Thornbird, or simply the Plain Thornbird; but it seems best to follow the established practice of giving the same vernacular name to a whole species. In Venezuela the thornbird is called the guaití. It has a discontinuous range in northern Venezuela and northeastern Colombia, northern Peru, and through Bolivia, north-central Argentina, and northern Paraguay, thence in a wide belt across Brazil to the Atlantic coast. Birds of warm lowlands, thornbirds ascend to about 3,000 feet (900 m) in Venezuela, where their nests were abundant in the botanic garden in Caracas.

Rufous-fronted Thornbirds are sedentary and appear to be permanent residents wherever they are found; established pairs probably never wander more than a few hundred yards from the nests in which they sleep throughout the year. These structures are rarely, if ever, in closed woodland but hang from trees that stand alone or in small groves in pastures, along roadsides and fence rows, and around farmhouses amid shade trees and shrubbery. Their nests are rarely far from

Habitat of Rufous-fronted Thornbird in northern Venezuela. They foraged in thickets on the hillsides and nested in trees beside the pasture.

thickets or dense, concealing vegetation through which the birds forage in pairs or family groups that seldom exceed six or seven individuals. When their nest tree stands in the midst of a cultivated field or clean pasture, the birds fly rapidly over the open spaces to the nearest thicket, where they hunt mainly on the ground, beneath bushes and weeds for insects and spiders. When fallen leaves cover the soil in the dry season, thornbirds often push their foreparts beneath the litter, or disappear into a loose accumulation of it, to emerge on the other side. I did not see them push or toss leaves aside, as many ground-foragers do. A pair feeding nestlings close by a house often hunted in a pile of rotting leaves raked up from the garden and crept through heaps of brush.

Occasionally, thornbirds climb up to investigate masses of dead leaves lodged in bushes and vine tangles, from which they pull out leaves and drop them. Although their nests often hang high, thornbirds rarely forage more than a few yards above the ground. Sometimes they hunt near Pale-breasted Spinetails, Buff-throated Wrens, or Southern House-Wrens, whose methods of foraging are similar, except that house-wrens often walk over the ground with alternately advancing feet, whereas the thornbirds hop with their feet together.

Although Rufous-fronted Thornbirds live in closest intimacy with members of their own family, they are strongly territorial and resist intrusion by members of other families. I did not try to trace territorial boundaries that often ran through thickets scarcely penetrable by man. During the five months that I lived with them, I witnessed only one clash between families. On a May morning, loud singing drew my attention to half a dozen or more thornbirds in the undergrowth of a patch of light woods, about midway between two nests, each occupied by six grown birds. All were flitting excitedly about, and sometimes one bird mildly chased another. Once two confronted each other momentarily, without coming to grips. Soon the two families drifted apart.

Thornbirds are seldom long silent. Their frequent outbursts of loud, ringing song, no less than their spells of quieter twittering, suggest a cheerful, contented nature. The song is a series of rapidly repeated similar notes that always sounds bright and joyous and at its best is beautiful, especially when the rather sharp notes that introduce a series merge into lower, mellower tones. Then it reminded me of Scaly-throated Leaftossers' songs, my favorites among ovenbirds' recitals. This loudest of the thornbirds' utterances also serves as a call to communicate with a distant mate. When close together, the partners often duet. Resting side by side on top of their nest, stretching upward until their bodies are almost vertical, tilting their heads skyward until at times their bills point straight upward, while their downward-directed tails beat time with their notes, they pour out their bright tones in unison. Although the voices of the sexes differ little, sometimes the notes of one, probably the male, sound stronger. Thornbirds sing not only on top of their nests but often within them, in the dim dawn light before they emerge in the morning, after they retire in the evening, or while one incubates or broods nestlings. The song of one family often stimulates members of a neighboring family to raise their voices.

Thornbirds frequently twitter with a rapid, continuous flow of weak notes that are sometimes squeaky. For minutes at a time I have heard twittering from a nest into which several had retired for the night, and again before they flew out at dawn. Thornbirds also twitter much while they build nests or attend their eggs or young. A duet by a mated pair may end with a twitter. Thornbirds twitter when two or

Rufous-fronted Thornbird (Plain race) *Phacellodomus rufifrons* (sexes alike) at nest. Venezuela to eastern Brazil and northwestern Argentina

more are close together, seeming thereby to express contentment or mild, pleasant excitement.

Among thornbirds' shorter utterances are monosyllables that, according to circumstances, range from a slight *tic* to a loud, sharp *chip*. These notes are strongest and most penetrating when a snake, a cat, a human, or some other potential enemy approaches their eggs or young. In the evening twilight, slight *tic*s issuing from a neighboring thicket often announce that a thornbird is about to fly up to its dormitory. In this situation, as when danger is apparent, the *tic*s seem to express anxiety, felt by a bird preparing to emerge from a sheltering thicket for a flight through the open air, exposed to attack by raptorial birds.

Nests

Unlike many other birds, thornbirds do not try to hide their nests; on the contrary, they choose for them exposed sites where the large structures hang free, untouched by surrounding branches. A dangling vine serves well for a nest's attachment. Occasionally the birds build amid a small colony of Yellow-rumped Caciques or Crested Oropendolas in an isolated tree with open branching, their bulky constructions contrasting with the more slender woven pouches of their neighbors. Although they prefer a tree standing alone in a field or beside a road, if none is available they may build at the edge of a grove or thicket, always on the more exposed side. The nests that I saw in Venezuela ranged from seven to about seventy-five feet (2 to 22 m) above the ground. Both of the extremes were in the same locality, on the open llanos in the state of Cojedes. Neither the lowest nor another a foot higher held eggs or nestlings.

The nests that most often catch the attention of the hurried traveler—large structures dangling from vertically descending, often leafless branches—give a false impression of the kinds of sites that thornbirds commonly select. Usually they start to build at or near the end of a slender, leafy, more or less horizontal branch with lateral branchlets

Rufous-fronted Thornbirds' nest on a horizontal branch

to prevent the nest's slipping off, at the outside of the tree's crown, or in the clear space beneath it. Unless the supporting branch is exceptionally stout, it sinks under the growing mass of sticks until the nest hangs vertically, while the foliage drops off or is removed by the birds, increasing its exposure. Rarely, a nest is built around the slender, upright trunk of a young tree.

Except the lining, thornbirds' nests are composed wholly of sticks, often twice as long as the bird who carries them. The longest stick that I found in a nest measured twenty-one inches (53 cm). Those over a foot (30 cm) in length are frequent, but many are only a few inches long. The stoutest ones are about a quarter of an inch (6 mm) thick—about as thick as a pencil. South of Lake Valencia, where rainfall was abundant during the wet season and most of the woody plants lacked thorns, the majority of the sticks in thornbirds' nests were thornless. Probably in more arid regions, where vegetation bristles with spines, our bird uses enough thorny twigs to justify its name. When Betsy Thomas offered building thornbirds equal numbers of thorny and thornless twigs of the same lengths, the birds preferred those with thorns.

Thornbirds sometimes try to break twigs from trees, occasionally with a few woodpeckerlike pecks to loosen them; but unless the piece is decayed, it will not yield to tugs of their slender bills. If not transferred from an earlier nest of the same pair, most of their sticks are loose ones gathered on or near the ground. Grasping a single twig in its bill, at or near the point of balance, the bird hops and flits upward through the nearest shrubs and trees, where possible to a point near or above the level of the nest, to which it flies on a horizontal or slightly inclined course across the intervening open space. The heavier the piece, the more the bird seeks the aid of conveniently situated branches to raise it gradually to the nest, resting here and there on the way. Often, however, the stick is borne upward by a fairly long and straight ascending flight. Sometimes the bird's rapidly beating wings are barely able to raise it aloft, and occasionally it is borne downward by the weight of its burden. Although thornbirds pass so much time in thickets where long flights are not needed, they can fly swiftly and straight.

After it has chosen a stick, the thornbird shows an indomitable will to carry it to its destination. It displays considerable skill in maneuvering the clumsy piece through obstructing branches, but now and

Rufous-fronted Thornbirds' many-chambered nest about seven feet (2 m) long, dangling from a liana on the llanos of Venezuela

then the stick is knocked from the bird's bill. A builder carrying a long, branched twig lost its balance when the piece struck an obstruction; but the bird's grip on the stick was so tenacious that bird and burden fell four yards to the ground together. Losing no time, the bird carried the piece to the rough, leaning trunk of the nest tree, up which it crept and hopped with its heavy load to the supporting branch, then flitted along this bough to the nest. Unlike the Firewood-gatherers described by Hudson, thornbirds frequently retrieve pieces that fall from the nest. I have seen them drop straight downward twenty-five feet or more to the ground in pursuit of a falling stick, then return to the nest by a circuitous course.

Nearly always one finds two birds cooperating closely at their work, and occasionally they have a helper or two. Building proceeds with much singing, by means of which the partners keep in touch and encourage one another while separately they search for sticks. When they meet at the nest, they duet and twitter together. Although their task is strenuous, they seem to enjoy it. Often, after alighting on the nest with a long stick laboriously raised to it, the builder continues for a short while to stand holding it in its slightly elevated bill, sometimes waving it around, in what appears a foolish, abstracted attitude. I surmised that the bird felt about its burden much as I did about the heavy stepladder that I had carried a long distance across the fields to a thornbirds' nest; although eager to be relieved of the load, setting it down was so awkward an operation that I sometimes stood holding it briefly after I had reached my destination.

After its short rest, the bird proceeds to fit the new piece into the nest. Placing the stick appears not to be preceded by a survey of the structure leading to a decision as to just where the latest contribution is needed. On the contrary, the thornbird holds the stick near the middle, with one end lower than the other, and thrusts it sideward while it continues to move over the nest. Often it repeats these apparently random movements a number of times before the lower end of the stick slips in between those already present. Sometimes the new piece is promptly accommodated in the fabric; at other times minutes are spent with a recalcitrant stick that does not seem to fit anywhere and may finally be laid loosely on top of the others.

While pushing a new twig into the nest, the builder vibrates or jerks its head, thereby keeping the stick's end slightly agitated until it encounters an interspace into which it can slip, and facilitating its passage through the maze of interlaced twigs. If the new piece fits too loosely, the bird may pull it out and continue to poke it sideways as before. After finally placing a new stick, the builder often seizes in turn the projecting ends of a number of others, testing their stability and pushing them deeper into the fabric if they are loose. Instead of promptly flying off for another stick, the bird may devote five or ten minutes to arranging materials already present, sometimes removing loose pieces from the bottom of the nest and inserting them at the top. By such continued testing and rearranging, the fabric is made strong.

One day I watched a thornbird trying to pull up a very long stick that had slipped through the bottom of a newly started nest but was prevented from falling by a fork at its upper end. With its bill, the bird drew up the stick as far as it could, but as soon as it released its hold for another lower on the stick, the piece slipped down until stopped by the fork. The bird tried this a dozen times, sometimes pulling the stick almost halfway out; but always the stubborn object fell back after the bird had lifted it as far upward as it could stretch and needed a lower hold to complete the operation. In the end, the thornbird flew away leaving the stick dangling below the nest. By using a foot to retain the stick while it secured a new grip with its bill, the bird might easily have solved its problem. Likewise, it might have drawn out the stick with its mate's cooperation. But the manipulation of a stick is always the task of a single bird.

At nests occupied by only a mated pair, only two birds build; but when more than two individuals lodged together, more than two might participate in enlarging their nest or building a new one nearby. At several nests I was certain that three members of a group, probably parents with their grown young, brought sticks and placed them in the structure. On one occasion, three individuals carried sticks to a nest under construction, while a fourth laid its twig on the old nest nearby. If my birds had been individually recognizable, I probably would have found more builders. For nearly six hours on five days, Betsy Thomas watched a family build a new nest to replace their damaged structure. The adult male brought 112 sticks; his mate, 69; while two of their three four-month-old offspring brought 17. However, the young birds contributed chiefly short pieces of nest lining. Moreover, 11 sticks were brought by unidentified individuals. Juveniles also assist in routine maintenance of nests.

When building a new nest or actively enlarging an old one, thornbirds begin work after an interval devoted to breakfasting at daybreak. In the cool early morning they start with great energy, but they appear soon to tire of their strenuous task; their trips to the nest with sticks become more widely spaced, and by the middle of the forenoon they usually rest from toil. Thus, two birds working on a recently begun nest brought 31 contributions from 6:35 to 7:35, 12 from 8:30 to 9:00, but only one from 9:00 to 9:30. At another nest, three birds brought material 38 times from 7:20 to 8:05, 12 times from 8:05 to 8:25, and 3 times from 8:25 to 8:45. After this concentrated building in the early morning, thornbirds may work sporadically at their nest at almost any hour of the day; and in the evening, before they retire for the night, they may build actively for a brief interval.

While building, thornbirds are often bothered by foliage that clusters around their growing nests, impeding the arrangement of their long sticks. They may push the disturbing leaves outward, only to have them spring back to their original positions after they are released. More often I have seen them try to tear away the offending foliage, usually with little success, for their bills are ill fitted for such work. Sometimes a thornbird reaches far up to grasp a leaf with a foot while it tugs at it with its bill. One bird clung to a large mango leaf with both feet while pulling it with its bill. Occasionally the bird pecks the recalcitrant leaf. Although such thick leaves as those of the mango tree are obdurate to the thornbirds' efforts, by dint of great persistence they do succeed in detaching small leaflets of compound leaves, or in tearing pieces from the edges of leaves. In removing some of the leaves that cluster around their nests, thornbirds differ from most arboreal birds, which seek concealing foliage. The removal of foliage to increase exposure is known chiefly in birds of which the males display for extended periods at a particular spot, such as certain manakins and birds of paradise.

If, while seeking sticks, thornbirds find some soft or flexible material, they may bring pieces for the lining, even to a nest that is hardly more than an open platform. I did not see them concentrate on lining their nest, as many birds do; they bring appropriate materials as they find them, not only while building but also frequently while incubating their eggs, and even occasionally while feeding nestlings. A family of thornbirds tore apart an old covered nest of a Great Kiskadee to thicken the lining of their own bulky nest, which had long been finished. Although thornbirds bring only one stick at a time, they may carry several pieces of flexible lining.

As building proceeds, the sides of the platform of sticks with which the nest begins are built up faster than the center, converting it into a shallow bowl and then a deeper cup. Then the walls contract inward, until the hollow is roofed over, becoming a nearly spherical chamber. Instead of proceeding to thatch this chamber with broad pieces of material, as spinetails (*Synallaxis*) do, the thornbirds continue to bring sticks and build a second, similar chamber above the first. The first surge of building rarely subsides until this upper chamber is covered with at least a few sticks, and sometimes it persists until a third chamber is begun atop the second.

Since thornbirds are indeterminate builders, adding to their structures at all seasons, while they incubate and even occasionally while they feed nestlings, it seems incorrect to say that a nest is ever completed. However, after the second chamber is at least loosely covered, some pairs relax their efforts, and the nest may be considered as temporarily finished.

In April, one pair took only ten or twelve days to build a two-chambered nest, with the upper compartment scantily roofed, after which they rested for five weeks before they began to lay. Later in the season, in July, another pair started to lay about eighteen days after they began their nest, when their upper chamber, although well lined, was still an open cup without a vestige of a roof. Yet this second pair, who built more slowly, transferred much of their material from their older nest only two yards (2 m) away; whereas the first pair had no such convenient quarry and perforce sought their twigs at a greater distance.

Two-chambered nests, which are the smallest in which, as a rule, one finds thornbirds sleeping or breeding, are about 15 or 16 inches high and from 9 to 14 inches in diameter (38 or 41 by 23 to 36 cm), not counting the ends of sticks that on all sides protrude far beyond the main mass, giving the nest a bristly, unkempt aspect. The globular chambers that they contain are about 4½ to 5 inches (11 to 13 cm) in diameter. The entrances to these rooms take various shapes. That of the lowest room, in which the brood is usually reared, may be an upwardly directed passageway through the sticks, which here bulge out farther than on the other sides of the nest. This hallway may be five or

Two nests of Rufous-fronted Thornbirds (the largest masses, far left and right center) in a colony of thirteen active nests of Yellow-rumped Caciques on the llanos of Venezuela

six inches (13 or 15 cm) long by about two inches (5 cm) in diameter. Sometimes it is shaped like the spout of a teakettle, or it may dilate inward to form a sort of vestibule or antechamber in front of the main chamber. Very often the entranceway bends in one direction or another, so that one cannot look straight down it and see what is inside the nest; but occasionally it is straighter, facilitating inspection. Although typically the external opening of the lowest chamber is near the top of a two-chambered nest, if the supporting branch has sunk far downward under the structure's weight, rotating its long axis through nearly ninety degrees, the doorway may be at the side, near what has now become the bottom of the nest. In an old nest with a number of compartments, the entranceways may take various directions, some leading downward to a chamber, others upward, and yet others horizontally; but as a rule all open in the same side of the nest. Each of the compartments has its own opening to the outside, with no internal communication between them; partitions of crisscrossing sticks separate them.

The chambers are lined on the bottom with almost any soft or flexible material that the thornbirds find. The lining of one nest was largely strips of fibrous bark; that of another, in the midst of a pasture, fibrous pieces of decaying stems and leaf-sheaths from the tall grasses. Thin, curving pieces of material are preferred to wide ones; but in one nest I found broad flakes of inner bark, apparently from a nearby woodpile, and a piece of decaying wood five inches long by nearly an inch broad (13 by 2.2 cm), as well as some small shriveled leaflets. Nests near human habitations have more diverse contents, such as scraps of cellophane, pieces of plastic bags (in one case six inches [15 cm] square), brightly colored candy wrappers, tinfoil, paper, and similar rubbish, as well as chicken feathers, vegetable fibers, scraps of wood, and strips of bark. The presence of lining in a chamber is not an indication that it has been, or will be, occupied by a brood. While incubating, the birds may deposit at least as much new lining in an upper chamber as in that where their eggs rest.

From time to time, the thornbirds add a new chamber to the top of their nest, or to the higher side of one on an inclined rather than a vertical branch, until it becomes an enormous mass, all out of proportion to its diminutive builders. They are so strongly attached to their nests that they might continue all their lives to enlarge them, but a limit is often set by what the supporting branch will bear. It may break under the growing weight, or as seems more frequently to happen, an angle or curvature of the branch makes further upward building impracticable. It was doubtless no accident that the tallest nest that I saw was built around a long, slender, vertically hanging liana, which seemed to invite the thornbirds to build up and up indefinitely. This inaccessible nest, on the llanos, was estimated to be seven feet (2 m) high and to contain eight or nine compartments. Nests three or four feet high, with four or five chambers, are not uncommon. Usually the chambers are in a single series, one above another; but an exceptionally broad nest contained one compartment in the bottom and two, side by side, at the top.

Despite their weight, thornbirds' nests seem rarely to fall unless the branch that supports them breaks from the tree. They do not become detached from the branch because they are built around it; it is firmly embedded in the wall, usually at the rear. In spite of the simple technique used in their construction and the lack of cement or other binding material, the wall of each chamber is surprisingly strong. To make an aperture large enough to insert a small mirror or electric bulb for viewing eggs or nestlings, I forced a stout, pointed stick into the fabric, which I could hardly open with my fingers. After each inspection,

I worked twigs into the gap to close it. Unlike spinetails, the thornbirds appeared not to notice what I had done.

Nests of Rufous-fronted Thornbirds are noteworthy chiefly for their size. Made of a single kind of material, lined with almost anything soft or flexible that the birds can find, they lack the elegance of the smaller, single-chambered nests of spinetails that use carefully selected materials for different parts. Unlike spinetails, thornbirds do not roof their chambers to keep them dry; the top, like the walls and floor, is covered with sticks. Probably little rain seeps into the lower compartments of very large nests; but in a nest with only two chambers, the loosely covered upper one must act as a basin to catch rainwater, which pours through the thin floor into the lower chamber where the birds sleep and rear their brood. Thornbirds appear to care little for dryness; I repeatedly failed to find them sheltering in their nests from a daytime shower, and an evening rain hardly advanced their hour for retiring. They may begin to sleep in an unfinished nest that is still a roofless cup. A family whose nest tree was cut down retired at nightfall into the hollow end of a rotting stub that afforded some concealment but no protection from the downpours frequent at that season. Although thornbirds inhabit some of the drier parts of the tropics rather than rain-forested regions, the wet season when they breed is not without hard rains.

The large nests of thornbirds remind us of the many-chambered structures of the Social Weaver of southern Africa, the Monk Parakeet of southern South America, and the Palmchat of Hispaniola, all of which build compound nests in which a number of pairs or families sleep and nest, each in its own compartment. However, the strong territoriality of Rufous-fronted Thornbirds prevents their nests from becoming avian apartment houses. No matter how many rooms it may have, each nest is the abode of a single breeding pair, often with their grown, nonbreeding offspring, plus at times a few surreptitious interlopers, whose presence is resisted by the family.

Why, then, do thornbirds start with two-chambered nests, to which at intervals they add more compartments, always working upward? The complexity of these structures may baffle predators. A snake or small mammal advancing along the supporting branch would first reach the upper rooms, which are not used for breeding unless the lowest chamber is damaged or occupied by an intruding bird of a different species. Finding these rooms empty, it might abandon its search for prey, and in any case the delay would give parents and fledged young time to escape. That the complexity of these structures does make it difficult to find the eggs I can attest from personal experience. Balancing myself precariously on a high, self-supported ladder that restricted my movements, I searched for eggs in the upper chamber of a nest and the antechamber of the lower one that I mistook for the lower compartment itself. Then, one day, after I had been examining the nest for several minutes, a thornbird flew out past my face, evidently having come from some part of the bulky structure that had escaped my scrutiny. Thereupon, I made a hole in the side, inserted my light and mirror, and saw two eggs that the bird had been incubating—my first thornbird's eggs.

Further to mislead searchers for their eggs—ornithologists or predators—parent thornbirds enter the unoccupied chambers of their mansions with confusing frequency. One may need to watch long to learn, from the movements of the birds, just where their progeny lie. Nevertheless, despite their complexity, strength, and relative inaccessibility, these nests are not invulnerable; seven of the nine of which I learned the outcome were pillaged, probably by arboreal snakes that crept in through the doorway, as I noticed no gap in the wall such as mammals might make. One pair lost two successive broods from the same nest before they built another nearby. In the multichambered nest, older offspring can sleep in rooms other than that where the parents incubate their eggs and brood their nestlings; but a smaller nest with a less permeable roof would, it seems, provide drier and more comfortable bedrooms. Whatever may be the advantages that have promoted the evolution of these big nests, it is evident that they are an extreme manifestation of a propensity widespread in the ovenbird family—that of building for its own sake, as an outlet for excess energy, a pastime, or a satisfying occupation.

Sedentary thornbirds are strongly attached to their chosen homesite. As long as they can, they add new chambers to their old nest instead of starting another. If this nest falls, they build a new one as close to the old site as they can. One would suppose that, when their nest is invaded by birds as aggressive as a pair of Troupials, they would move to a safe distance, but instead of this they build a new nest close to that still occupied by these dangerous intruders. When they have been robbed of eggs or nestlings, they may lay again in the pillaged structure, often in the same chamber as before, or they build their replacement nest nearby. These masses of coarse sticks last a long while, even after they have been abandoned, and one often notices two or three hanging prominently in the same tree. A family of thornbirds, finding it no longer practicable to add a new chamber to the top of their nest in the usual manner, started another below the old one at the end of the same branch, which had grown longer since the earlier structure was begun. When the new nest was finished, the projecting ends of its sticks overlapped those of the old nest above it.

Other thornbirds that build hanging nests are the Greater, the Little, the Streaked-fronted, and the Freckled-breasted, but none of these birds makes nests as large as those of the Rufous-fronted. Most thornbirds live in southern South America, from Bolivia and Brazil to central Argentina.

Sleeping Habits

On an evening in mid-March, the day after I arrived at the hacienda La Araguata where I studied thornbirds, I watched a three-chambered nest that hung thirteen feet (4 m) up in a tree that grew in a fence line close by the farm buildings. Nearby, the diesel engine that drove the electric generator was chugging loudly. At 6:50, when the light was growing dim, I heard repeated sharp notes. Soon several thornbirds appeared low in the weedy fence line about a hundred feet (30 m) from the nest. Passing through the pungent fumes from the engine, seeming not to be troubled either by its noise or by my unconcealed presence, they advanced, staying among the herbage near the ground. When near the nest, they flew up to it, either directly or by way of a neighboring tree. Four entered through two doorways, and after a few minutes two more joined them. It was then nearly dark. Although in the four days that I had already spent in Venezuela I had noticed many thornbirds' nests along the highways, these were the first thornbirds that I saw, with the exception of two that flew from a roadside nest as we sped by.

At 6:25 next morning, as it was growing light, I watched the six thornbirds fly from their nest. After leaving, they vanished down the fence line, uttering a few sharp notes. Soon they sang amid the dense vegetation on the bank of a neighboring stream.

It was then the height of a long, severe dry season, which was to continue well into May. As I became familiar with the farm, I found enough thornbirds to suggest that they were among the most abundant birds in the area, but they foraged so obscurely that I rarely saw them except while I watched them enter their nests at nightfall or leave at dawn. Because they darted in or out very rapidly, sometimes I needed to count two or three times to be sure how many occupied a nest. By May 1 I had investigated fourteen nests within about a mile and a half (2 km) of my residence. Although rarely I counted seven birds at a nest, I could never repeat these counts. No nest that I studied had more than six regular lodgers; three had this number. One nest was occupied nightly by five thornbirds; at another I sometimes found five and sometimes four. Each of three nests sheltered three sleepers. Six nests were occupied by pairs of birds. In May and June, when many thornbirds were breeding but none had yet fledged young, I investigated eight additional nests, finding in each no more than two birds past the nestling stage. Thus, before the number of grown birds was augmented by the year's offspring, fourteen of twenty-two nests, or 64 percent, were occupied by only a mated pair. These couples without grown companions were apparently young birds nesting for the first time or older ones who in the preceding year had not succeeded in rearing progeny that survived.

Usually the thornbirds retired late in the evening and arose early in the morning; but I found a good deal of variation between nests, and even between the several occupants of the same nest. The birds who slept near the noisy electric plant went to rest very late, when little daylight remained, possibly because of this and other disturbances. But even farther afield, with no sounds save those of the natural world, families differed in the hours they slept. The six birds who slept in nest 11 flew forth in the dim light of dawn, so early that after watching them leave I could reach nest 17, a hundred yards away, some minutes before the six sleepers left this nest. Similarly, these birds retired earlier than their neighbors of nest 11. And even at a single nest with three or more occupants, the first might enter ten or fifteen minutes before the last. The latest arrivals, who might enter in the dusk when I could hardly see them, were, at least in some cases, intruders rather than members of the family.

When only two thornbirds occupied a nest, I always found them sleeping in the same room, whether or not they had eggs or nestlings. When three or more were present, it was often difficult to learn how they distributed themselves for sleeping among several available rooms. Often they would enter by different doorways, but then they might shift from chamber to chamber; and these restless movements would continue until the light had become so dim that I could hardly distinguish the dark birds as they crept rapidly over their dark nest from one entrance to another. In the growing obscurity, I was not sure that I had witnessed the last of these changes. Similarly, in the dim light before the birds flew down at dawn, they would often shift from chamber to chamber; and I was not certain that this activity had not started before the day had become bright enough to reveal it to me. Moreover, if I gave too much attention to how the birds distributed themselves among the rooms, I was likely to miscount the number that entered or left the nest as a whole.

Despite these difficulties, repeated watching convinced me that the thornbirds were not consistent in occupying the chambers available to them. Sometimes one would force its way in with others who resisted its intrusion, when it might without opposition have entered another compartment of the same nest. Parents with eggs or nestlings often tried to exclude their older offspring from the brood chamber, not always successfully. Despite occasional fluctuations in the number of sleepers in a nest, through April and May families remained intact. Twelve nests that in March and April had a total of forty-four members had in late May the same number, not counting nestlings. In June and July, when many pairs were incubating or feeding nestlings, yearlings tended to leave the family abode to seek mates and establish homes of their own. Apparently, they would not breed until the following year, the second since they hatched. Betsy Thomas's banded thornbirds remained with their parents for as long as sixteen months.

In addition to the united family, nests are sometimes occupied at night by other thornbirds who never become integrated with the family. These intruders are sometimes birds who have just lost their nests and impose themselves upon unwilling neighbors until they can build new shelters for themselves, but in other cases they appear to be immatures, or older birds without mates and territories. A nest with several chambers was occupied by three birds who were evidently a mated pair with a yearling and two others unrelated to them. The first three slept together in the middle chamber. Many minutes after they retired, a fourth individual appeared in a neighboring tree, calling attention to itself by low, sharp notes expressive of anxiety; it seemed to be nervous about approaching the nest. Finally, it flew up to the structure, only to dart away a moment later. Then, when the light had become dim, this bird and another went to the nest at the unusually late hour of 7:07 P.M. Now excited twittering came from the dark structure. I could distinguish the birds only when they were silhouetted against the sky as one or more of them flitted restlessly over the nest. Soon movement and voices ceased as the birds settled down for the night. I could not tell whether all five were together or in different rooms.

For at least two months, these five thornbirds lodged in the same nest without becoming intimate. When, to my intense annoyance, the nest tree was cut down for fence posts, the three members of the family took refuge in the hollow top of a neighboring stub. When it was nearly dark, one of the interlopers, after an initial repulse, pushed into a high nest in another territory, where a pair of thornbirds and a Piratic Flycatcher were incubating. I could not learn what had happened to the other interloper. This incident helped explain some of the fluctuations in the number of sleepers in a nest that I sometimes noticed.

Later, I discovered a case of intrusion following the loss of a nest. After a family of four moved into a new nest built on the same nearly horizontal branch of a roadside *Erythrina* tree that supported its predecessor, the overladen bough snapped off. The birds promptly started another nest on a neighboring bough of the same tree. While this nest was only a platform, or at best an open bowl, I watched to learn where the birds without a roof would sleep. After sunset I found two of them bringing sticks to the incipient nest. Soon they settled on the platform, as though to roost there, amid the foliage that clustered around it; but after staying a few minutes, they flew toward a sandbox tree about 250 feet (75 m) away, where a single pair had a small new nest. Unsettled, these two returned and left again. The other two birds who had slept in the *Erythrina* tree before the branch fell failed to appear that evening.

While I watched the *Erythrina* tree, my wife, Pamela, watching at the sandbox tree, saw four birds arrive, one by one. Two entered the nest that evidently belonged to them, but the third met resistance at the doorway. One of the first two grappled with the third, and they fell into the bushes below the nest. After they arose, the third bird persisted in trying to force its way in with the first two, causing much singing and twittering, much going in and out of the nest. When I joined Pamela at the sandbox tree, the intruder from the *Erythrina* tree was resting in the doorway, with its tail sticking out into the light of the rising Moon. Finally, it pushed inside, causing more twittering to issue from the hanging nest. Meanwhile, we continued to hear sharp *chips* coming from the low, tangled vegetation below the structure. They did not cease until Moon and stars were shining brightly. We waited until half past seven without seeing this bird fly up to the nest; we could hardly have missed it because it would have been silhouetted against the moonlit sky. Apparently, too weary to face the opposition of the neighboring pair, it passed the night amid the foliage.

On the following evening, Pamela watched the *Erythrina* tree while I watched the nest in the sandbox tree. She reported that two birds came to the unfinished nest, left, came again, but soon flew toward the sandbox tree. Then a single bird returned, rested on the nest, departed, and finally came to pass the night on the open platform.

Meanwhile, at 6:47 P.M., I saw the resident pair enter the nest in the sandbox tree, followed by the usual loud singing and contented twittering. Nearly a quarter of an hour later, another thornbird flew up to the nest, but instead of promptly entering the lower chamber with the first two, it remained for some minutes on the top, or perhaps on the farther side—at least, I lost sight of it. Presently it approached the doorway of the lower chamber, but apparently denied admittance by the unseen birds within, it withdrew a short distance. Again and again it tried to enter only to be repulsed. Soon it adopted an expedient to meet this situation. After each ineffectual attempt to enter, it turned around and stood with its tail inward, reminding me of a Red-crowned Woodpecker entering tailfirst a hole of which, apparently, it was slightly suspicious. This about-face and presentation of its tail to, I suppose, the pecks of an unseen bird within happened many times, while the almost full Moon grew brighter and more stars shone out. Finally, at nearly half past seven, the intruder pushed in at least far enough to pass from view and stayed.

On the following evenings, the intruder entered the nest in the sandbox tree fifteen or twenty minutes after the resident pair, who had apparently become more or less reconciled to its presence and seemed no longer to try strenuously to exclude it. A lone bird rested on the new nest in the *Erythrina* tree, which continued to grow slowly. By mid-July this nest had been covered over and two birds lodged in it. To my surprise, the sandbox-tree nest was still occupied at night by three thornbirds, one of whom left at daybreak well in advance of the other two. Had the bird who all this while continued to roost in the unfinished nest in the *Erythrina* tree acquired a new partner? Or had the bird who forced its way into the sandbox-tree nest rejoined its mate in the *Erythrina* tree, and had another homeless thornbird found lodging in the sandbox tree?

By delaying until it is almost dark, when the rightful occupants have become drowsy and can hardly see the intruder or distinguish it from other members of their family, homeless thornbirds become unwanted guests in nests of established pairs, and may continue for months to impose themselves upon their reluctant hosts.

Incubation and Care of Nestlings

Like other birds that forage much on the ground, the thornbirds nested later than birds that find their food in trees or the air. They waited until returning rains had soaked the ground litter and quickened the small creatures that inhabit it. An exception was a pair established in a low meadow near a stream, where the soil remained moister, and invertebrates more abundant, than in areas better drained. This pair laid at the end of April; and in another inaccessible nest, also near a brook, the birds were incubating early in May. Although in early May a few light showers fell and humidity increased, the rainy season in the year (1966) I made my observations did not become well established until after the middle of the month. In the second half of May frequent, copious showers greened hillsides that had long been brown. Now the thornbirds began to lay more freely, and by late May and June many pairs were incubating. Each of the three clutches that I could reach consisted of three plain white eggs, always lying in the lowest chamber, unless this had been preempted by an intruder of another species.

Both sexes incubated. I gave most attention to incubation by a pair of which one member was tailless in April and remained in this condition for the next four months. I believe this was the female; but the partners took such equal shares in nest attendance that the designation of the sexes is of little importance. Both had bare incubation patches on the abdomen, as I saw clearly when they preened while standing in the doorway with their breasts toward me. As at other nests, both parents slept every night in the chamber with the eggs. I watched this nest through two mornings and two afternoons. Taking the four half days together, during 25½ hours the bird with a tail took 41 sessions totaling 662 minutes; its tailless partner, 39 sessions totaling 700 minutes. The sessions of the former ranged from 4 to 42 minutes and averaged 16.1 minutes; those of the latter varied from 1 to 54 minutes and averaged 17.9 minutes. The longest interval when neither was in the brood chamber was 12 minutes. The longest interval between changeovers was 79 minutes, when the tailed bird was in charge of the nest. During the next-to-longest interval, 54 minutes, the tailless one was in charge. During the first morning and the afternoon of the next day, the eggs were attended for 89.8 percent of the time. A week later, one or the other bird was in the nest for 88.7 percent of a morning and the next afternoon.

Although on warm, sunny afternoons each partner left the eggs once, twice, or even thrice during its period in charge of the nest, it rarely left the nest itself. The nest was unattended for only two minutes, while the tailed bird chased a trespassing thornbird. When not inside the brood chamber, the bird in charge stood in the doorway preening, singing, or simply gazing out. Or it would emerge and go over the nest, pulling up falling sticks or pushing in loose ones, sometimes shifting a piece of material from one place to another, as spinetails do. When coming to take their turns at incubation, or on special trips, the thornbirds brought a stick, more often shreds of snakeskin or other pliable materials for the lining. These were deposited not only in the lowest compartment with the eggs but often in the middle chamber of this nest with three rooms. An item was sometimes thrown out or carried to a distance.

Throughout the day, these thornbirds often sang while standing in the doorway, resting on top of the nest, on a stick projecting from its side, on a nearby branch, or even while inside the nest. The mate might answer from a distance; or they might duet, especially as one replaced the other on the eggs. Sometimes the tailless partner twittered while the other sang loudly, suggesting that it was the female.

Once, when both partners were in front of their nest, neighboring birds of other kinds sounded an alarm, possibly because a raptor that I did not see was passing by. Both thornbirds instantly dived into their nest, where they stayed in silence for about two minutes. Evidently, they felt safe from aerial attack inside their strong-walled castle. This was the only time that I saw thornbirds take refuge in their nest when alarmed.

At another nest I watched through a mostly cloudy forenoon with intermittent fine drizzles. I could not distinguish by appearance the two partners who shared incubation, but one sang in a deeper voice than the other. Their 13 sessions of incubation ranged from 1 to 59 minutes and averaged 28.9 minutes. The eggs were unattended for only 3 intervals totaling 10 minutes. This pair were in the chamber with the eggs for 97.4 percent of 6½ hours, exceptionally high constancy for ovenbirds. Like the first pair, they often came with soft materials for the nest's lining. Twice during the morning, a flock of Groove-billed Anis, coming to eat the arillate seeds of a tree of the anona family that supported the nest, jumped all around the structure, shaking it, even alighting upon it, but on neither occasion did the incubating bird so much as look out. More than most birds, Rufous-fronted Thornbirds permit birds of other species to come near their nests, probably because their eggs and nestlings are out of sight and well protected, but they promptly drive away thornbirds not of their own family.

Thornbirds are reluctant to leave their eggs unguarded. If the mate is tardy in coming to take its turn at incubation, the partner at the nest sings loudly to recall it to its duty. One morning I watched a thornbird, whose mate was neglectful, stand in its doorway and sing over and over. After many minutes of this loud calling, its notes became distinctly higher and weaker, as though it was becoming hoarse. Finally, failing to obtain a response, it flew down into a thicket and continued to sing in its altered voice.

My prolonged watches were at nests inhabited by only the breeding pair. At a nest with six grown occupants where incubation had just started or was about to begin, some members of the family entered the brood chamber at nightfall in a manner which suggested that a parent within was trying to keep them out. One who had just entered emerged slowly and reluctantly, as though driven from within. After clinging awhile beside the doorway, it entered again, and again a bird, this one or another, was forced slowly outward. This happened over and over, until finally five or six individuals remained to sleep in the brood chamber. Unhappily, Troupials broke up this nesting before I could make further observations.

I could determine the incubation period at only one nest. Sixteen or seventeen days elapsed from the laying of the last egg to the hatching of the last nestling.

Thornbirds hatch with the pink skin, sparse gray down, and tightly closed eyes usual in the ovenbird family. The interior of the mouth is yellow instead of black, as in the adults. Both parents continue to sleep in the chamber with the nestlings, as they did with the eggs, and older brothers and sisters sometimes accompany them. I passed a morning watching a nest with three nestlings ten and eleven days old. No grown birds other than the parents occupied this nest. The first parent left in the dim light at 6:07 A.M., and the first meal was brought to the nestlings at 6:15. In the next fifteen minutes, the young were fed 11 times by both parents. The number of meals brought each hour from 6:07 to 12:07 was 31, 20, 31, 11, 10, 18. In the six hours of the morning, the three nestlings were fed 121 times, or at the rate of 6.7 times per nestling per hour.

Two days later, Pamela and I took turns watching this nest from 12:07 P.M. until the second parent retired for the night at 7:06. The last meal was brought in the fading light at 6:55. From 12:07, the number of meals brought each hour was 14, 22, 19, 19, 17, 22, 21. In nearly seven hours of the afternoon, the three nestlings were fed 134 times, or at the rate of 6.4 times per nestling per hour. Taking the morning and afternoon records together, in a day of nearly thirteen hours the nestlings received 255 meals.

When these three nestlings were twenty and twenty-one days old, both parents left the nest at 6:10 A.M. and brought the first food at 6:22. In successive hours from 6:10 onward, the number of meals served was 18, 10, 7, 16, 23, 25. The total of 99 meals in six hours was substantially less than the 121 received in this interval ten days earlier. Such a reduction in the rate of feeding is not unusual with nestlings who linger in the nest for days after they have passed the stage of most rapid growth and, already well feathered, use less energy to keep warm.

Although the parents of this brood were difficult to distinguish, they appeared to take equal shares in attending their nestlings. As far as I could see, on each visit to the nest they brought a single item, held conspicuously in the tip of the bill. Except an occasional round object that might have been a berry but was more probably an egg case of some sort, the nestlings were nourished wholly with insects and other invertebrates, which were quite small at first and rarely large even after the young were feathered. Small brown pupal cases were frequently brought. Other items that I recognized were green caterpillars, small grasshoppers, small or middle-sized moths, and rarely a spider. Occasionally, the parents came with a stick or lining for the nest instead of food.

The unfinished upper chamber of this new nest had developed a hole in the rear wall in addition to the doorway in front. Bringing food, the parents nearly always passed through this chamber from back to front, then climbed down the front of the nest to the lower compartment where their nestlings rested. To leave, they reversed their course, traversing the upper chamber from front to rear before they flew away. This indirect course appeared to be an expression of the thornbirds' habit of creeping through closed spaces rather than a ruse to mislead predators seeking their brood. When they carried away a dropping in the bill, the parents omitted this passage through the upper chamber. They promptly removed the shells from which the nestlings hatched, and kept the nest clean at all times.

When I found yearling thornbirds helping their parents to build, I fully expected that they would feed their younger siblings. Since my birds were not individually recognizable, I could verify this only by seeing three bring food simultaneously, or two while a third brooded, neither of which I did. Neither did Betsy Thomas, who lived for years near thornbirds and banded them, report such feeding. In other families of birds, nonbreeding individuals help to feed more often than they help to build; but ovenbirds are such persistent builders that with them we might expect the reverse to be true.

During the forenoon of July 1, when the three above-mentioned nestlings were ten and eleven days old and their plumage was beginning to expand, a parent stayed in the nest, apparently brooding, for 18 intervals, ranging from 1 to 13 minutes and totaling 70 minutes. The morning was clear except for about an hour when the sky was clouded, and from 11:00 to noon the sun shone hotly. Two days later, when these nestlings were fairly well covered by their rapidly expanding feathers, they were brooded, during seven hours of the afternoon, for 11 periods ranging from 2 to 38 minutes and totaling 170 minutes. Sunshine alternated with showers that were mostly brief and light. Even when the nestlings were sixteen and seventeen days old and well feathered, during two hours of a sunny afternoon they were brooded for intervals of 6, 11, and 4 minutes. Thereafter, I noticed no more daytime brooding. By night the parents were probably in close contact with the nestlings as long as they remained in the nest, even if they did not always cover them.

The Young after Fledging

When these nestlings were eighteen and nineteen days old and no longer brooded by day, the parents, who had formerly gone completely inside to deliver food, sometimes fed with the end of the tail projecting from the doorway. The young birds greeted the arrival of a meal with fine, rapidly repeated, insectlike notes. On the following day, I first heard the nestlings give a weak version of the adults' song. When they were twenty-one days old, they sometimes advanced far enough into the antechamber to take a meal from a parent who remained outside. Occasionally a nestling revealed itself in the entrance while it was fed—hitherto, the young had always remained out of sight. One of them came out, turned around, dropped its excreta over the side of the nest, and promptly reentered. Now the young birds often repeated their weak song, and sometimes they uttered an infantile version of the *chip*. I did not hear nestlings twitter.

At this nest, as at another, the nestlings left on the day after I first saw them expose themselves briefly in front of their doorway. The parents seemed excited by their impending departure, for early in the morning of their last day in the nest one of them gave an old feather to a nestling, as though it were food; it was apparently not swallowed. Soon after this occurred, a young bird, following a parent who had just delivered a meal, emerged from the chamber and crawled around the side of the nest, then promptly reversed its course and reentered. By eight o'clock one of the fledglings was resting in the top of a neighboring rose-apple tree. A tuft of nestling down still adhering to its head, together with its brighter, fresher plumage, its shorter bill, and the yellow corners of its mouth distinguished it from its parents, who were preening nearby. The young bird was about as large as the adults, and its tail appeared as long as theirs. Its two siblings came out later that same day. The two older fledglings were twenty-two days old; the other about one day younger.

Earlier, on June 5, I had watched the departure of another brood, reared in an inaccessible nest into which I had first seen the parents carry food on May 15. Song floated from the high nest as I arrived before sunrise. Six grown birds had slept in this nest, four or five of them in the brood chamber with the fledglings, one or two in an upper compartment. After flying from the nest, these thornbirds called much and were obviously excited. About the time the last grown bird emerged, two fledglings appeared in the doorway. One crept out, climbed to the top of the nest, then returned to its natal chamber. Soon some of the adults returned without food, and stood beside the young in the doorway. The latter came out in front and withdrew into the nest again. Then, at 6:25 A.M., with no parental prompting that was evident to me, one fledgling launched forth, and a minute later the other followed. The three-week-old thornbirds flew well on a descending course to trees seventy-five and a hundred feet (23 and 30 m), respectively, from the nest. The adults sang much after their departure. The fledglings soon vanished into low, dense vegetation, whence I heard their weak *tsips*, although I could no longer see them.

Late in the afternoon of June 5, I found the two newly emerged fledglings with grown birds, amid dense bushy growth along a fence between pastures, where they had gone early in the morning. Here they remained until, at 6:25 P.M., the parents led one of them upward through the branches of a mango tree to its summit. A parent flew to the nest, entered the brood chamber as though to inspect it, then promptly rejoined the others in the mango tree. Now the fledgling, accompanying both parents, easily flew to the nest tree, a distance of about a hundred feet (30 m), on a nearly level course. The adults went directly to the nest, but the young bird continued past it and alighted in a fork of the nest tree. From here it immediately flew back to the nest and without perplexity found and entered the lower chamber, where it had grown up. It was then 6:30; the summits of the western hills were still bathed in golden sunshine, but the valley where I watched lay in shadow.

After the entry of the fledgling, its parents continued to pass back and forth between the brood chamber and the upper chamber. Soon one flew down. After a while, the other parent, apparently hearing the weak calls of the second fledgling, who had stayed behind in the bushes, joined it there and led it to the top of the mango tree, whence it flew with its parents to the nest tree. It, too, continued past the nest to alight in a crotch, but with only a little more difficulty than the first had experienced, it gained the brood chamber, at 6:47. After a little more going in and out, the parents stayed in this chamber with the young. These two adults alone led the fledglings back to the nest; the other four grown birds who lodged there were still absent. When they arrived, about a quarter of an hour later, the first three went directly into the upper room, while the last joined the parents and fledglings below them, at 7:05. Including the two fledglings, this nest now sheltered eight thornbirds, the greatest number that I found lodging in any nest.

Six days later, I again watched this family retire in the evening. As before, the parents and fledglings ascended to the top of the mango tree, whence a parent and both young flew to the nest, which now the latter entered without first veering past it. A minute later the second parent arrived, and both busied themselves arranging sticks before they went in. Soon two more adults reached the nest and passed from chamber to chamber before they settled down. The other two grown birds who formerly lodged here failed to appear; I never, to my knowledge, saw them again. Probably these two were the parents' first brood of the preceding year, while the two nonbreeding adults who remained were the second brood. Thus, soon after the latest brood was fledged, the number of occupants of this nest was again reduced to six.

I last saw one of these fledglings receive food from a parent on June 17, when it had been out of the nest for twelve days and was about thirty-four days old—the age at which many small passerines can forage for themselves.

The first days in the open of the fledglings of certain woodpeckers, wrens, and other birds that sleep in dormitories are considerably shorter than those of their parents. They leave the nest late in the morning, retire early in the evening, and are fed by their elders after their return to the nest. But I never saw a thornbird feed a fledgling in the nest after its first flight. From the beginning, young thornbirds spent long days in the open. On the day they left their nest in an algarrobo tree, three young remained abroad with their parents until 6:45 P.M., when the family of five gathered in a small, acacialike tree near the nest. From this point a parent flew to the nest, followed by the others. All tried to enter at the same time, jamming the passage. Then some came out while others tried to push in, causing much confusion. By 6:54, however, all had settled down inside, except one parent, who had flown off with a dropping. Five minutes later this adult returned and entered. Thereafter, none left, although from time to time one peered through the doorway as daylight waned.

At 6:04 next morning, the parents flew from the nest. They sang much, and the young still inside joined in with their weaker voices. A parent reentered, then left, the brood chamber; but no food was brought. When the three fledglings left the nest ten minutes later, much singing greeted their departure. Six days after this, the parents and two surviving young left the nest together at 6:10. The third fledgling was killed by a cat.

The Second Brood

Early in July, the pair of thornbirds whose nestlings I had watched leave the nest on the morning of June 5 gave indications that they would breed again. Late in the forenoon of July 5, I found two adults and a juvenile on the nest in which the latter had been hatched. On the following day, toward noon, two adults and a juvenile were again at the nest, and one of the former brought a stick. One of them bit the young bird mildly, but it did not retreat. On the next morning, July 7, I saw only the two adults at the nest. A week later they were incubating in this inaccessible structure, in the chamber where their first brood had been reared, and also bringing new materials, most of which they deposited, not there, but in the room above. One of the parents, coming to the nest with a stick, was followed by the two juveniles of the first brood. Thrice the adult flew at the young birds without pressing the attack. They retreated only a few inches, then climbed unmolested over the nest.

These parents became increasingly unfriendly to their progeny who continued to lodge in their nest. When I arrived at sunset on July 20, a juvenile, recognized by its whiter throat and breast, was resting on top, preening and at times lightly adjusting a stick. A parent was in the lower chamber, incubating, and from time to time coming to the doorway to look out. The other parent brought a long stick. The juvenile, then the adult, flew down.

When, a little later, a thornbird came to the entrance of the brood chamber, a parent darted out and chased it down into the bushes. Then one parent entered the brood chamber and another bird, probably the other parent, went into the upper chamber. As the light grew dim, three more thornbirds arrived and rested on sticks projecting from the bottom of the nest. Whenever one of these latecomers ascended to the doorway of the brood chamber, the parent sallied forth and attacked it, making it retreat, then returned inside. This happened over and over. As twilight deepened, the parent in the upper chamber joined its mate in the room with the eggs. Then two of the birds who had been waiting at the bottom of the nest cautiously climbed up and entered the upper chamber; whereupon a parent emerged from the brood chamber, ascended to the upper chamber, and drove them out. They stayed close to the doorway, and after the parent's prompt return to the brood chamber, reentered the room from which they had been evicted. The third of the late arrivals climbed up and joined them. Before it was quite dark, all had settled down, three in the upper compartment and two in the brood chamber.

At break of the following day, I watched this nest again. The five sleepers became active late, when the day was growing bright. While waiting to fly down, one or more of them passed repeatedly from chamber to chamber. They also emerged only to turn around and reenter the same room. Now I detected no discord among the five.

The six thornbirds that in March and April slept in this and each of two other nests were apparently a mated pair with the young of two broods raised in the preceding nesting season. If parents habitually show such antagonism to their older offspring as I witnessed on the evening of July 20, how can a family of six be built up, when no more than three young are raised in a brood? No other thornbirds that I watched displayed such persistent enmity to grown progeny who shared their nest. Nevertheless, their effort to exclude the other sleepers was ineffectual, and the parents' bad temper did not persist overnight. Thornbirds who lack nests of their own are amazingly pertinacious in entering the nests of other thornbirds, and this persistence is probably greatest when the nest is the familiar home in which they grew up. Doubtless, the antagonism of these parents was associated with the onset of incubation of the second brood and would wane after the eggs hatched, so that by the end of the breeding season they would again dwell peaceably with whatever offspring remained with them.

20 *Enemies, Nest Success, & Survival*

We have no reason to suppose that adult and juvenile ovenbirds are more immune to predation by raptors, small mammals, and other carnivores than are birds of other families that share their habitats, but I have never seen a grown ovenbird fall victim to a predator, and published records of this are few. On the Falkland Islands a Blackish Cinclodes was, strangely, seen to be killed by an Austral Thrush, and the skull of another was found in a pellet of a Short-eared Owl. Humans have been the major enemy of ovenbirds, as of so many other creatures. Fortunately for the ovenbirds, their lack of colorful plumage and melodious song has saved them from the wicked cage-bird trade, and they are probably too lean and elusive to be attractive as food to aboriginal hunters. Too many have been killed to fill the specimen trays of natural history museums; but humans have chiefly injured ovenbirds by destroying on a vast and accelerated scale the rain forests where many species dwell. With ovenbirds that inhabit secondary vegetation, humankind's more complex relations will be considered in the following chapter.

Whereas grown ovenbirds appear to have relatively few enemies, other than habitat-destroying and habitat-poisoning humans, their nests have many. In Argentina, a young white-eared opossum preyed upon two broods of Rufous Horneros; Fraga found it in one of these nests with the partly eaten corpses of the mother and her nestlings. A Black-chested Buzzard-Eagle tore open one horneros' nest and removed the nestlings. Horneros' nests on fence posts around corrals are accidentally wrecked by horses and cattle and more deliberately by humans.

The spinetails' nests that I have studied in Central America are more vulnerable to predation than are those of horneros in Argentina. Although I have never witnessed an act of predation, the nest-robbers are evidently of two kinds. Those that make round holes, one and a half to three inches (38 to 76 mm) in diameter, in the walls of nests are most probably small mammals, possibly large, carnivorous bats, one of which, *Vampyrum spectrum*, is known to eat birds as large as anis. The predators that extract eggs or young, possibly with the brooding parent, through the tunnel, with no structural damage to the nest, are doubtless snakes, and perhaps also mammals no bigger than rats, which abound in thickets where spinetails nest.

Spinetails unconcerned in the presence of humans react strongly to serpents. As I was ending a morning vigil at a Slaty Spinetails' nest with nestlings a few days old, a parent emerged from the tunnel and flew to the ground, where it began to repeat sharp monosyllables similar to the usual call note but stronger and more insistent. Promptly joining it, the other parent called in the same manner. Soon I noticed a slender green snake, about two feet (60 cm) long, stretched out on a tangle of vines near the ground. The snake slid deeper into the massed vegetation, and the birds, still scolding sharply, hopped back and forth through the maze where it lurked, coming very close to it, if not actually pecking it. Knowing these green tree snakes to be nest-robbers, I removed it; but a few days later the nest had been pillaged, probably by another snake, as the structure was intact. Very different was the behavior of this same pair of Slaty Spinetails toward a gray, crested lizard, nearly a foot in length including its long tail, that for over an hour clung upright to a twig less than a yard from their nest, apparently unnoticed by them. Likewise, these birds paid no attention to the shiny little lizards called "lucias" that crawled over their nest.

In Venezuela, a pair of Rufous-fronted Thornbirds' sharp *chip*s of alarm drew my attention to a black-and-yellow "tigra," a snake over six feet (2m) long, resting in a rose-apple tree whose boughs met those of the algarrobo tree where their nest hung. For at least an hour and a half, the thornbirds, neglecting their nestlings, continued to hop close around the serpent, repeating their alarm notes. When the snake stretched out, they bravely pecked the tail of a creature so much larger than themselves; as though a man should, bare-handed, attack a great shark. When the snake coiled up, they did not touch it, as far as I saw, but they often approached within a few inches of the thick part of its body. After a while, the serpent aroused itself and crept toward the thornbirds' nest in the most direct line. Knowing *Spilotes pullatus* to be an insatiable pillager of birds' nests, I seized a stick and knocked it to the ground, over which it raced too swiftly to be killed. The thornbirds continued for many minutes to look for it in the trees near their nest, then resumed feeding their hungry nestlings.

The volubility of nestling ovenbirds of at least one species is disastrous to them. Hudson gave a vivid account of how the voices of young Hudson's Canasteros, called by him Hudson's Spinetails (the species having been named for him by Philip Lutley Sclater), betrayed them to the Chimango Caracara. This little canastero creeps like a mouse through and beneath the pampas grasses. It hides its domed nest of woven grasses, with a small orifice in the side, so well beneath the densest part of a cardoon thistle that Hudson searched every day for a whole season without finding one. But the high-pitched tones of the nestlings conversing together between parental visits made them so easy for Chimangos to find that these carrion-eaters fed their own young on little else. This reminds us once again how capricious natural selection, which over long ages adapts some organisms so finely to their circumstances, may fail to accomplish something so seemingly simple as silencing nestlings' voices responsible for their wholesale destruction.

Covered nests and woodpeckers' holes are coveted by many birds that lack the skill to make such shelters for their own broods. These tenants that pay no rent are of two kinds: those that by force or cunning wrest nests from their builders and rightful owners, and those that move in after their makers have abandoned them. Only the former can be regarded as enemies, but since some of the tenants may be at one time usurpers and at another time peaceable inheritors, we may consider both categories together. An abode so commodious and safe as the Rufous Horneros' ovens of clay has many claimants. One of the most persistent of these is the Brown-chested Martin, which Hudson called the Tree Martin because he never saw it alight elsewhere than in trees, never on the ground or buildings, as other swallows often do. In Buenos Aires Province he found it breeding only in horneros' nests, and described in vivid detail its efforts to wrest them from their makers. A migrant, it breeds later than the resident horneros and often finds ovens from which they have already brought forth their young, in which case no conflict arises. But if for any reason the horneros' nesting is delayed, as by the need to replace a lost brood, the martins, now ready to breed, become aggressive and try to dispossess the builders of their still-occupied structures. They approach closer and closer to the coveted nest, only to be driven away by the larger horneros until, becoming bolder, they alight at the doorway. Now the desperate defenders fiercely attack the resolute invaders; again and again they clutch together and drop to the ground. After days of struggle, the invaders often win. With large feathers they cover the horneros' eggs, and upon them lay their own. When the horneros' young have hatched and their parents' zeal is at highest pitch, they frequently succeed in holding the aggressors aloof until their young have fledged, when the martins move in before the chamber has grown cold. Their efforts to capture a clay oven are frequently complicated by those of persistent and aggressive Saffron Finches, who often carry feathers and straws into nests still occupied by eggs or young of the horneros.

House Sparrows, introduced from Europe, are a more recent threat to horneros' domestic tranquility. Of forty-six of their ovens examined by Joanna Burger in Santa Fe Province, seven, or 16 percent, were occupied by these aliens. Instead of coming to grips with the bigger ovenbirds, they repeatedly approach an oven only to be chased away by the owners, until the latter, wearying of repulsing the pestilent intruders, abandon their nest. Among other birds that occupy the ovens, with or without struggling to obtain them, are Tufted Tit-Spinetails, Cattle Flycatchers, Chopi Blackbirds, Southern House-Wrens, and parrotlets (*Forpus* sp.). Some of these same birds, including Cattle Flycatchers and Saffron Finches, also occupy Firewood-gatherers' commodious mansions of sticks, as do Bay-winged Cowbirds and White-rumped Swallows, not without contests among these intruders for exclusive possession. Strangely, Pale-legged Horneros, capable of building respectable clay ovens for themselves, occasionally prefer to lay their eggs in nests of the Rufous Cacholote or the Rufous-fronted Thornbird.

Rufous-fronted Thornbirds' nests with two or more separate chambers offer house room to a variety of tenants, sometimes simultaneously with the owners. Blue-gray Tanagers not infrequently build open cups in thornbirds' nests, choosing old chambers with gaps in their walls. The similar Sayaca Tanager has the same habit. Cattle Flycatchers often occupy thornbirds' hanging structures. An aggressive tenant is the Piratic Flycatcher, which captures enclosed nests of a wide diversity of birds by the expedient of throwing out, at opportune moments, the eggs or, less often, the nestlings. This nearly always causes builders of simple nests to abandon them, but thornbirds may continue to occupy a house of several rooms with the intruders. A pair of Piratic Flycatchers reared two broods in a high thornbirds' nest still inhabited by the latter. At one time the Pirates and their hosts fed nestlings in separate chambers of the same edifice, one above the other. The thornbirds failed to rear their brood. Cooperatively breeding Striped-backed Wrens often occupy thornbirds' nests, whether after evicting the builders or otherwise, I do not know. Different birds find firm support for their nests on top of the thornbirds' massive, pendant structures. A pair of Great Kiskadees built their domed nest of straws on the roof of one of these constructions. After their departure, a pair of little Thick-billed Euphonias made their nest inside the kiskadees' capacious chamber.

Of all the intruders into Rufous-fronted Thornbirds' nests, the most destructive is the truculent Troupial, an atypical oriole that appears incapable of weaving a nest, as other orioles so skillfully do. When it breeds in a crowded colony of Yellow-rumped Caciques, it not only destroys the eggs or young in the hanging pouch that it selects for its own eggs, but also violently evicts the owners of a number of surrounding nests to increase the safety of its own progeny; a predator that finds several pouches empty may abandon its search for eggs or nestlings before it reaches those of the Troupial.

Where Troupials coexist with Rufous-fronted Thornbirds, as in northern Venezuela, their nests are the pirates' preferred abode. When a pair of these colorful, melodious rascals claimed a large thornbirds' nest with five rooms in which six thornbirds slept, they tore larger entrances in the sides of the two lowest chambers. The lowest of all was chosen as a sleeping room by the female Troupial, who repeatedly repelled attempts of her partner to share it with her. He slept in the next lowest. Gradually the invaders pulled sticks from a zone around the middle of the nest, until it had a constricted, hourglass appearance, a sign that a thornbirds' nest is, or was, occupied by Troupials. Sometimes this alteration so weakens the structure that the part below the constriction, where the Troupials breed, breaks away and falls. For several weeks, the six thornbirds continued to lodge in their upper compartments, above the sleeping intruders. During this interval, I noticed no conflict between the two kinds of birds. The thornbirds were building a new nest in the same small tree, into which the six moved as soon as it was habitable.

By early June the female Troupial was incubating three eggs in the room where she had long lodged. They rested upon a thick bed of finely shredded materials that she had arranged over the coarser lining originally applied by the thornbirds. Dissatisfied with his first bedroom, the male Troupial tore a large gap in the side of the lowest chamber of the new nest, destroyed the eggs that a thornbird had laid there, and passed the night in it. When this happened, the thornbirds resumed sleeping in their first nest, in the chamber above the incubating female Troupial. Unwillingly on their part, the thornbirds and the male Troupial had exchanged domiciles.

After their eggs appeared, the Troupials became fiercely antagonistic toward the industrious little birds who had provided lodgings and a nest site for them. One morning I watched a Troupial drive a thornbird from tree to tree, then across twenty-five yards of open pasture into a thicket, where the smaller bird escaped through vine tangles that impeded its pursuer. Rarely have I seen one bird chase another so long, with such fierce persistence. These attacks continued until the thornbirds could occupy a new nest, their third, that they built for themselves, screened from the first two by a large, umbrageous mango tree.

About the time the young Troupials fledged, I noticed a large gap in the side of the lower room in a two-chambered nest, belonging to a single pair of thornbirds, about 250 feet (76 m) from the nest that the larger family had lost to the Troupials. The nestlings that this pair had been feeding had vanished. In the evening, I watched the bereaved parents trying to mend the gap in their wall, but when a Troupial arrived it pulled out more sticks before it entered. Then the thornbirds retired into their upper chamber. To rear a brood of two, the Troupials had opened two nests built successively by one family of thornbirds and one nest of a neighboring pair, destroying one set of thornbirds' eggs and one brood of nestlings, probably three of each. Fortunately, these brilliant rogues, the national bird of Venezuela, are in the north of that country much less abundant than the obscure, laborious birds whose nests they invade. Although thornbirds and Troupials may sleep in different rooms of the same nest, it is doubtful whether the former could rear a brood in a structure occupied by the latter.

Although the thornbirds that I studied did not in my presence try to defend their nests from the much bigger aggressors, Betsy Thomas saw two families of thornbirds try to reoccupy chambers that Troupials had pirated from them. On each occasion, the Troupials pushed into the rooms where the thornbirds had been sleeping and evicted them. One thornbird chased by a Troupial after eviction, was not seen again, possibly having been killed by the pursuer.

Not only birds but also a diversity of other creatures take shelter in the snug, enclosed nests of ovenbirds, among them rats, snakes, lizards, social wasps, ants, termites, spiders, and other arthropods. A colony of small, black, stingless bees called "aragres" established itself in a Slaty Spinetails' nest, binding the sticks together with a blackish, resinous cement and making a funnel-like entrance that projected from the wall of twigs. Beside the Río Napo in Peru, I peeped into an oven of clay, probably built by the Pale-legged Hornero, and found a solitary frog.

Ovenbirds' well-enclosed nests are no more exempt from the intrusion of parasitic cowbirds than are open nests. Herbert Friedmann and Lloyd Kiff listed twenty-five species as hosts of the Shiny Cowbird, widespread in South America. Four of these parasitized species—the Rufous Hornero, Firewood-gatherer, Short-billed Canastero, and Olive Spinetail—are reported to have reared cowbirds. Rufous Horneros sometimes eject cowbirds' eggs, as do a number of birds in North America, where the widespread cowbird is the Brown-headed.

A more insidious parasite is the Striped Cuckoo, which intrudes its blue eggs into the covered nests of other birds, especially ovenbirds, wrens, and certain finches. In Argentina alone its known hosts include nine species of ovenbirds—spinetails, canasteros, and thornbirds. It was long supposed that the twelve-inch (30 cm) cuckoo could not enter the well-enclosed structures made by birds half its size without tearing a hole in the wall, which the victims would promptly close. However, Hugh Land photographed a slender Striped Cuckoo entering one of the best-enclosed of all ovenbirds' nests, that of the Rufous-breasted Spinetail, through the narrow tunnel, then emerging without difficulty. Like Brown-headed Cowbirds, Striped Cuckoos lay their eggs early in the morning, while their victims are breakfasting. Although hosts of cowbirds frequently rear one or more of their own young along with one or two of the fosterlings, this cannot happen with the cuckoo because its nestling, like that of parasitic honeyguides of Africa, bears on the tips of its mandibles sharp hooks with which it murders any nestling that lies beside it.

The Rufous Horneros studied by Rosendo Fraga in Buenos Aires Province, Argentina, nested with high success. Of 115 eggs laid in 33 clutches, 83, or 72.2 percent, yielded fledglings, which compares favorably with the reproductive success of hole-nesting birds in the North Temperate Zone. The main cause of nestling mortality appeared to be starvation, especially of the youngest member of broods of four, who hatched more than twenty-four hours after the first-born and was usually smaller and lighter than its siblings. Although the horneros in this population frequently laid four eggs, they were often unable to rear more than three young. These horneros owed their success largely to the paucity of predators, especially tree-climbing snakes, in the parkland and tree plantation where they were established. In other parts of Argentina, horneros nested with much poorer results. Not only did Fraga's horneros enjoy high nesting success, they also lived fairly long. The minimum average annual survival rate of territory-holders was 71.4 percent, which is higher than that of most small birds in the North Temperate Zone (for which we have the most data) but lower than that of tropical birds.

In their big nests of sticks, the Nores's Brown Cacholotes did not do as well as Fraga's Rufous Horneros in their ovens of clay. Of 177 eggs laid in 67 nests, 144 (81.3 percent) hatched. Of the 144 nestlings, 105 (72.9 percent) fledged, or 59.3 percent of the eggs yielded young who survived to leave the nest. Ten eggs failed to hatch, 8 were taken by white-eared opossums, 7 by rats, and 6 by various birds. Of the nestlings, 12 died in the nest, and predators took 24. On four occasions, nests pillaged by white-eared opossums were occupied by them. Older, experienced cacholotes nested with significantly greater success than first-time breeders. Yearlings with previous experience in breeding did better than those of the same age without experience.

Slaty Spinetails that I studied in snake-infested thickets and coffee plantations in Costa Rica suffered much higher losses than did these Argentine ovenbirds. Of 13 nests found before incubation began and followed through to the end, only 3, or 23 percent, produced at least one fledgling. This is poor success even for nests in clearings, thickets, and light second-growth woods in the humid tropics. Of 756 nests of 23 species in the valley where these spinetails lived, 277, or 37 percent, yielded one or more fledglings. In this same region, Orange-billed Nightingale-Thrushes, whose habitats broadly overlap with those of Slaty Spinetails, lay two eggs in mossy open cups built in low thickets, fields of maize, light woods, and similar situations. Of 27 of their nests of known outcome, 17, or 63 percent, were successful by the same criterion—nearly three times the success of the spinetails. The apparently safer nests of Rufous-fronted Thornbirds hanging high in trees fared no better; only two of nine nests, or 22.2 percent, yielded flying young.

Why do spinetails, thornbirds, and other ovenbirds spend so much time and energy building mansions of sticks when they might rear as many, or more, young in simpler, more readily constructed nests? Perhaps other ovenbirds in other regions rear their families more safely in their enclosed nests; to answer our question convincingly we need many more studies in diverse lands. In any case, by renesting, perhaps repeatedly, in a long breeding season, Slaty Spinetails and Rufous-fronted Thornbirds manage to maintain their numbers in favorable habitats despite the loss of more than three quarters of their broods. In the following chapter, we shall return to this problem.

21 *Ovenbirds & Humans*

The ovenbird most familiar to the human inhabitants of the lands where it abounds, and most famous abroad, is undoubtedly the Rufous Hornero with its ovens of clay. Their fondness for the industrious bird that builds its massive structures on their houses and fence posts or in their dooryard trees is expressed by the names they give it: João de Barro—John of the Mud-puddles, or John Clay—in Brazil; Alonzo García or Alonzito, the affectionate diminutive, in Paraguay and adjoining parts of Argentina. Since a bird that receives a Christian name should act like a good Christian, the hornero is reputed to rest from its labors on Sundays and all church holidays. This bird that might be regarded as a symbol of peaceful industry has become Argentina's national emblem, and the republic's ornithological society call its journal *El Hornero*.

More than most birds, ovenbirds are fearless of humans. I have already told how Rufous-breasted and Slaty spinetails continued their domestic tasks while I stood little more than arm's length away. Striped-breasted and Yellow-throated spinetails are equally confiding. In Chile, Rufous-banded and Buff-breasted miners attend their nests only two yards from a person watching. On Tierra del Fuego, Thorn-tailed Rayaditos flit within reach of a human who passes through their territory, incessantly repeating *chip*s that attract others of their kind, until one is accompanied by two or three pairs of the inquisitive little birds. In the high Andes of Peru, the Bar-winged Cinclodes is one of the most familiar and fearless of birds, frequenting villages and dooryards, so tame that a person can approach within two or three steps before it rises to fly a few feet ahead or merely moves aside.

As we would expect, birds that are tame on the continent are even more confiding on islands. On the Falklands, the Blackish Cinclodes, there called the Tussoc Bird, is so fearless of humans that it will closely approach or even alight upon one who sits quietly. These dusky birds fly over the water to alight on an approaching dinghy, and settle on the rails of a steamship passing near the shore. They enter houses and at mealtime come to the dining table for tidbits. In a quite different setting, a Scaly-throated Leaftosser, who for half an hour made the leaves fly from the forest floor while I stood close above it, at the conclusion of its strenuous activity rested on a log almost at my feet. In the same rain forest, the Buff-throated Automolus is less approachable.

How human activity may affect ovenbirds is well illustrated by the history of the Slaty Spinetail in El General, at the head of the Térraba Valley, where I write. Before humans' arrival, this bird, which avoids the rain forest that thickly covered the region, must have been restricted to streamside thickets and small natural openings, if present at all. When Indians and, centuries later, people of European origin arrived and practiced shifting subsistence agriculture, resting fields soon became overgrown with the low, lush vegetation in which Slaty Spinetails thrive. When I came here in the mid-1930s, and for several decades thereafter, they were abundant in thickets, coffee plantations, and bushy pastures threaded by winding cowpaths. I found more of their nests than I had time to study. After this, with increasing human settlement, agricultural practices changed radically; cash crops replaced subsistence farming. Small plantations of tall coffee shrubs shaded by high trees gave way to large plantations of low shrubs with scarcely any shade. Small patches of sugarcane, selectively cut throughout the year, were replaced by much larger cane fields, clear-cut and burned in the dry season. Pesticides, herbicides, and other poisons, rarely applied in the early days, were lavishly spread almost everywhere with little regard for consequences. Now I rarely see or hear Slaty Spinetails.

The local history of the Pale-breasted Spinetail is somewhat similar. Early in this century it was established in savannalike areas lower in the Térraba Valley, but a broad barrier of almost uninterrupted forest blocked its access to the more recent clearings in the upper valley. With the almost total destruction of this magnificent rain forest in the midcentury, the spinetail expanded its range up the valley. But soon after it reached our neighborhood, the use of biocides increased, and its sojourn here was brief.

The only instance of ovenbirds becoming troublesome to humans that has come to my attention was reported by Helmut Sick. In the southernmost Brazilian state of Rio Grande do Sul, Firewood-gatherers damage electric installations by building their big nests upon them. Wire and other metallic objects that the birds include in their nests cause short circuits and fires. The offending nests are removed and burnt to prevent the prompt use of their materials in the construction of new nests. Rufous Horneros and Monk Parakeets are responsible for similar problems.

Inhabitants of tall forests, dense thickets, scrublands, stony mountain slopes, or high Andean páramo and puna, almost wholly insectivorous, ovenbirds appear rarely to clash with humans' economic interests. Their importance to us is aesthetic, scientific, even, I venture to suggest, philosophical and moral. They pose questions fundamental to our understanding of nature. One of these, already mentioned, is why some ovenbirds, such as the Slaty Spinetail, expend great effort to build elaborate nests that, on presently available evidence, diminish rather than increase their reproduction. This is contrary to present evolutionary theory, which holds that organisms should maximize the number of their progeny. A possible explanation of this apparent anomaly is that the habit of building enclosed nests of sticks originated in situations very different from those in which we find many ovenbirds today, such as the more open lands of South America. As ovenbirds spread northward, nest styles that may have increased breeding success in the regions where they arose proved to be maladaptive in the more recently colonized regions because the large, conspicuous structures were too vulnerable to the predators that they attracted. Nevertheless, the birds continued to build them from ancestral habit—an example of evolutionary inertia.

Although this hypothesis sounds reasonable, I doubt that it is an adequate explanation. Probably we should seek a psychic factor. Watching spinetails and thornbirds build nests that appear needlessly big and elaborate, meticulously caring for the finished structures as few other birds do, lavishly furnishing them with soft, flexible materials, such as cast reptile skins, that apparently do not increase their immunity to predation or otherwise improve their adequacy for their

Double-banded Graytail *Xenerpestes minlosi* Sexes similar Colombia and eastern Panama

primary purpose of rearing young, I have been impressed by these birds' strong attachment to edifices they have so laboriously made. They appear to take pride in their housekeeping, and seem to *enjoy* building and maintaining their nests, which may be the reason why they make them so elaborate.

Here we encounter a problem that arises in other contexts of avian life. Do birds enjoy singing and hearing their own voices, as they appear most convincingly to do when they tirelessly repeat melodious songs, improvise, and mimic? Do they feel affection for the young they so faithfully attend? Do their cries when their broods are threatened or lost express genuine feeling? Are the more social species comforted by close association with companions with whom they forage, reciprocally preen, and sleep in contact?

These are questions beyond the purview of the science of ornithology. Unable to observe directly or to measure with instruments the psychic life of birds, it can make no irrefragable pronouncements about it. Evolutionists offer us a picture of organisms of all kinds bending all their efforts to multiply progeny, but they hardly even suggest a reason for such endeavor. Creatures appear to reproduce just for the sake of reproducing. But what could be more absurd than the multiplication of animals that find no joys or satisfactions in their lives, no gleams of pleasure; what would be lost to them if they became extinct? If one were to suggest that the purpose of evolution is to increase the worth of animate existence, he may be greeted with stony silence if not with contempt by the more doctrinaire of his colleagues. We are left with a cold, comfortless view of a relentlessly harsh process.

In this predicament, what should we do? Probably we should have more confidence in the soundness of our spontaneous sympathies and reasonable intuitions. They are more likely to be true revelations of reality than are many of the beliefs that religions propose to comfort us and give us hope; and like religious beliefs, they can profoundly affect our worldview and the tone of our lives. To believe that young animals enjoy their gambols, birds take pleasure in their singing, and nonhuman parents feel affection for their young makes evolution, and indeed the whole world process of which it is a phase, appear less coldly alien to us than when we hold that all such apparent manifestations of value in living are no more than behaviors that promote survival and reproduction for no more reason than to survive and reproduce in an endless chain of vacuity. Since ovenbirds give us reason to believe that they enjoy building their nests, they are not without philosophic importance.

Ovenbirds have great capacity for caring. They care for their young as well as other passerines—or better, when they lead their fledglings to sleep in the nest from which they have flown, as only a minority of birds do. Their continuing care for their structures is exceptional; even those superb weavers, the oropendolas, fail to care for their finished pouches, permitting them to fall from their high treetops when a little timely mending might save them from becoming detached. Caring is one of the most advanced manifestations of animal life; if human beings may be regarded as the highest of animals, it is, above all, because we have the greatest capacity for caring (very unequally developed among individuals), for our families, our fellows, our future, the Earth that supports us, and much more. To contemplate ovenbirds' care for the more elaborate of their nests is heartwarming. These versatile architects deserve to be better known.

Bibliography

Bosque, C., and M. Lentino. 1987. The nest, eggs, and young of the White-whiskered Spinetail (*Synallaxis* [*Poecilurus*] *candei*). *Wilson Bull.* 99:104– 106.

Burger, J. 1976. House Sparrows usurp hornero nests in Argentina. *Wilson Bull.* 88: 357–358.

Cawkell, E. M., and J. E. Hamilton. 1961. Birds of the Falkland Islands. *Ibis* 103a: 1–27.

Caziani, S. M., and J. J. Protomastro. 1991. Nest and eggs of the Stripe-backed Antbird (*Myrmorchilus strigilatus*). *Condor* 93:443–444.

De la Peña, M. R. 1979. *Aves de la Provincia de Santa Fe.* Provincia de Santa Fe [Argentina], Ministerio de Agricultura y Ganadería.

Dorst, J. 1957. The puya stands of the Peruvian high plateau as a bird habitat. *Ibis* 99:594–599.

Dyrcz, A. 1987. Observations at a nest of the Pale-legged Hornero in southeastern Peru. *J. Field Ornith.* 58:428–431.

Edwards, E. P., and R. B. Lea. 1955. Birds of the Monserrate Area, Chiapas, Mexico. *Condor* 57:31–54.

Erard, C. 1982. Le nid et la ponte de *Lipaugus vociferans,* Cotingidé, et de *Grallaria varia,* Formicariidé. *Alauda* 50:311–313.

Euler, C. 1867. Beiträge zur Naturgenschichte der Vögel Brasiliens. *Jour. f. Ornith.* 15:177–198, 217–233, 399–420.

ffrench, R. 1973. *A guide to the birds of Trinidad and Tobago.* Wynnewood, Penn.: Livingston Publishing Co.

Fraga, R. M. 1979. Helpers at the nest in passerines from Buenos Aires Province, Argentina. *Auk* 96:606–608.

———. 1980. The breeding of Rufous Horneros. *Condor* 82:58–68.

Friedmann, H., and L. F. Kiff. 1985. The parasitic cowbirds and their hosts. *Proc. Western Foundation Vert. Zool.* 2:226–302.

Goodall, J. D., A. W. Johnson, and R. A. Phillipi B. 1957. *Las aves de Chile: Su conocimiento y sus costumbres.* Buenos Aires: Platt Establecimientos Gráficos.

Gradwohl, J., and R. Greenberg. 1980. The formation of antwren flocks on Barro Colorado Island, Panama. *Auk* 97:385–395.

———. 1984. Search behavior of the Checker-throated Antwren foraging in aerial leaf litter. *Behav. Ecol. Sociobiol.* 15:281–285.

Graves, G. R., and G. Arango. 1988. Nest-site selection, nest, and eggs of the Stout-billed Cinclodes (*Cinclodes excelsior*), a high Andean furnariid. *Condor* 90:251–253.

Greenberg, R., and J. Gradwohl. 1980. Leaf surface specializations of birds and arthropods in a Panamanian forest. *Oecologia* (Berlin) 46:115–124.

———. 1983. Sexual roles in the Dot-winged Antwren (*Microrhopias quixensis*), a tropical forest passerine. *Auk* 100:920–925.

———. 1985. A comparative study of the social organization of antwrens on Barro Colorado Island, Panama. In *Neotropical Ornithology*, ed. P. A. Buckley, M. S. Foster, E. S. Morton, R. S. Ridgely, and F. G. Buckley, pp. 845–855. Ornith. Monogr. 36. Washington, D.C.: American Ornithologists' Union.

Haverschmidt, F. 1953. Notes on the life history of the Black-crested Antshrike in Surinam. *Wilson Bull.* 65:242–251.

———. 1968. *Birds of Surinam.* Wynnewood, Penn.: Livingston Publishing Co.

Hilty, S. L. 1975. Notes on a nest and behavior of the Chestnut-crowned Gnateater. *Condor* 77:513–514.

Hilty, S. L., and W. L. Brown. 1986. *A guide to the birds of Colombia.* Princeton, N.J.: Princeton Univ. Press.

Hudson, W. H. 1920. *Birds of La Plata.* London: J. M. Dent and Sons.

Johnson, R. A. 1953. Breeding notes on two Panamanian antbirds. *Auk* 70:494–496.

———. 1954. The behavior of birds attending army ant raids on Barro Colorado Island, Panama Canal Zone. *Proc. Linn. Soc. N.Y.* 63–65:41–70.

Lill, A., and R. P. ffrench. 1970. Nesting of the Plain Antvireo *Dysithamnus mentalis andrei* in Trinidad, West Indies. *Ibis* 112:267–268.

Marchant, S. 1960. The breeding of some s. w. Ecuadorian birds. *Ibis* 102:349–382.

Meyer de Schauensee, R. 1970. *A guide to the birds of South America.* Wynnewood, Penn.: Livingston Publishing Co.

Miller, A. H. 1963. Seasonal activity and ecology of the avifauna of an American equatorial cloud forest. *Univ. Calif. Publ. Zool.* 66:1–78.

Munn, C. A. 1985. Permanent canopy and understory flocks in Amazonia: Species composition and population density. In *Neotropical Ornithology*, ed. P. A. Buckley, M. S. Foster, E. S. Morton, R. S. Ridgely, and F. G. Buckley, pp. 683–710. Ornith. Monogr. 36. Washington, D.C.: American Ornithologists' Union.

Munn, C. A., and J. W. Terborgh. 1979. Multispecies territoriality in Neotropical foraging flocks. *Condor* 81:338–347.

Narosky, S., R. Fraga, and M. de la Peña. 1983. *Nidificación de las aves argentinas (Dendrocolaptidae y Furnariidae).* Buenos Aires: Asociación Ornitológica del Plata.

Nores, A. I., and M. Nores. 1994. Nest building and nesting behavior of the Brown Cacholote. *Wilson Bull.* 106:106–120.

Oniki, Y. 1971. Parental care and nesting in the Rufous-throated Antbird, *Gymnopithys rufigula,* in Amapá, Brazil. *Wilson Bull.* 83:347–351.

———. 1975. The behavior and ecology of Slaty Antshrikes (*Thamnophilus punctatus*) on Barro Colorado Island, Panama Canal Zone. *Anais Acad. brasil. Ciênc.* 47: 477–515.

Oniki, Y., and E. O. Willis. 1983. A study of breeding birds of the Belém Area, Brazil IV. Formicariidae to Pipridae. *Ciência e Cultura* 35:1325–1329.

Pearson, O. T. 1953. Use of caves by hummingbirds and other species at high altitudes in Peru. *Condor* 55:17–20.

Remsen, J. V., Jr. and T. A. Parker, III. 1984. Arboreal dead-leaf-searching birds of the Neotropics. *Condor* 86:36–41.

Rowley, J. S. 1984. Breeding records of land birds in Oaxaca, México. *Proc. Western Foundation Vert. Zool.* 2:74–221.

Schneirla, T. C. 1949. Army-ant life and behavior under dry season conditions. *Amer. Mus. Nat. Hist. Bull.* 94:1–81.

———. 1956. The army ants. *Smithsonian Report for 1955*, 379–406.

———. 1957. A comparison of species and genera in the ant subfamily Dorylinae with respect to functional patterns. *Insectes Sociaux* 4:259–298.

Schwartz, P. 1957. Observaciones sobre *Grallaricula ferrugineipectus. Bol. Soc. Venezolana de Cien. Nat.* 18:42–62.

Sick, H. 1984. *Ornitologia Brasileira.* Brasilia: Editora Universidade de Brasilia.

———. 1985. Gnateater. In *A dictionary of birds,* ed. B. Campbell and E. Lack. Calton, England: T. and A. D. Posner.

———. 1993. *Birds in Brazil,* trans. W. Belton. Princeton, N.J.: Princeton Univ. Press.

Skutch, A. F. 1946. Life histories of two Panamanian antbirds. *Condor* 48:16–28.

———. 1966. A breeding bird census and nesting success in Central America. *Ibis* 108:1–16.

———. 1967. *Life histories of Central American highland birds.* Publ. Nuttall Ornith. Club 7.

———. 1969a. *Life histories of Central American birds.* Vol. 3. Pacific Coast Avifauna 35. Berkeley, Calif.: Cooper Ornithological Society.

———. 1969b. A study of the Rufous-fronted Thornbird and associated birds. *Wilson Bull.* 81:5–43, 123–139.

———. 1972. *Studies of tropical American birds.* Publ. Nuttall Ornith. Club 10.

———. 1981. *New studies of tropical American birds.* Publ. Nuttall Ornith. Club 19.

———. 1985. Clutch size, nesting success, and predation on nests of Neotropical birds, reviewed. In *Neotropical Ornithology,* ed. P. A. Buckley, M. S. Foster, E. S. Morton, R. S. Ridgely, and F. G. Buckley, pp. 575–594. Ornith. Monogr. 36. Washington, D.C.: American Ornithologists' Union.

Snow, D. W., and A. Lill. 1974. Longevity records for some Neotropical land birds. *Condor* 76:262–267.

Snow, D. W., and B. K. Snow. 1964. Breeding seasons and annual cycles of Trinidad land-birds. *Zoologica* (N.Y. Zool. Soc.) 49:1–39.

Studer, A., and J. Vielliard. 1990. The nest of the Wing-banded Hornero *Furnarius figulus* in northeastern Brazil. *Ararajuba* 1:39–41.

Thomas, B. T. 1983. The Plain-fronted Thornbird: Nest construction, material choice, and nest defense behavior. *Wilson Bull.* 95:106–117.

Tostain, O., and J-L. Dujardin. 1988. Nesting of the Wing-banded Antbird and the Thrush-like Antpitta in French Guiana. *Condor* 90:236–239.

Vaurie, C. 1980. *Taxonomy and geographical distribution of the Furnariidae (Aves, Passeriformes).* Amer. Mus. Nat. Hist. *Bull.* 166:1–357.

Vehrencamp, S. L., F. G. Stiles, and J. W. Bradbury. 1977. Observations on the foraging behavior and avian prey of the Neotropical carnivorous bat *Vampyrum spectrum. J. Mammal.* 58:469–478.

Vuilleumier, F. 1967. Mixed species flocks in Patagonian forests, with remarks on interspecies flock formation. *Condor* 69:400–404.

———. 1969. Field notes on some birds of the Bolivian Andes. *Ibis* 111: 599–608.

Wetmore, A. 1972. *Birds of the Republic of Panama,* Part 3. *Passeriformes: Dendrocolaptidae (woodcreepers) to Oxyruncidae (sharpbills).* Washington, D.C.: Smithsonian Institution Press.

Wiedenfeld, D. A. 1982. A nest of the Pale-billed Antpitta (*Grallaria carrikeri*) with comparative remarks on antpitta nests. *Wilson Bull.* 94:580–582.

Wiley, R. H. 1971. Cooperative roles in mixed flocks of antwrens. (Formicariidae). *Auk* 88:881–892.

Willis, E. O. 1967. The behavior of Bicolored Antbirds. *Univ. Calif. Publ. Zool.* 79:1–132.

———. 1968a. Studies of the behavior of the Lunulated and Salvin's antbirds. *Condor* 70:128–148.

———. 1968b. Taxonomy and behavior of Pale-faced Antbirds. *Auk* 85: 253–264.

———. 1969. On the behavior of five species of *Rhegmatorhina,* ant-following antbirds of the Amazon basin. *Wilson Bull.* 81:363–395.

———. 1972a. *The behavior of Spotted Antbirds.* Amer. Ornith. Union, *Ornith. Monogr.* 10:1–162.

———. 1972b. Breeding of the White-plumed Antbird (*Pithys albifrons*). *Auk* 89: 192–193.

———. 1973a. The behavior of Ocellated Antbirds. *Smithsonian Contrib. Zool.* 144:1–57.

———. 1973b. Survival rates for visited and unvisited nests of Bicolored Antbirds. *Auk* 90:263–267.

———. 1979. Comportamento e Ecologia da Mae-de-Taota, *Phlegopsis nigromaculata* (D'Orbigny e Lafresnaye) (Aves, Formicariidae). *Rev. Brasil. Biol.* 39:117–159.

———. 1982. The behavior of Scale-backed Antbirds. *Wilson Bull.* 94:447– 462.

———. 1984. *Dysithamnus* and *Thamnomanes* (Aves, Formicariidae) as army ant followers. *Papêis Avulsos Zoologia* (São Paulo) 35:183–187.

———. 1985a. Antbirds. In *A dictionary of birds,* ed. B. Campbell and E. Lack. Calton, England: T. and A. D. Poyser.

———. 1985b. Antthrushes, antpittas, and gnateaters (Aves, Formicariidae) as army ant followers. *Rev. Brasil. Zool.* 2:443–448.

Willis, E. O., and E. Eisenmann. 1979. A revised list of birds of Barro Colorado Island, Panama. *Smithsonian Contrib. Zool.* 291:1–31.

Willis, E. O., and Y. Oniki. 1972. Ecology and nesting behavior of the Chestnut-backed Antbird (*Myrmeciza exsul*). *Condor* 74:87–98.

———. 1978. Birds and army ants. *Ann. Rev. Ecol. Syst.* 9:243–263.

———. 1988. Nesting of the Rusty-backed Antwren, *Formicivora rufa* (Wied, 1831) (Aves, Formicariidae). *Rev. Brasil. Biol.* 48:635–637.

Willis, E. O., Y. Oniki, and W. R. Silva. 1983. On the behavior of Rufous Gnateaters (*Conopophaga lineata,* Formicariidae). *Naturalia* (São Paulo) 8:67–83.

Young, A. M. 1971. Roosting of a Spotted Antbird (Formicariidae: *Hylophylax naevioides*) in Costa Rica. *Condor* 73:367–368.

Index

Boldface page numbers indicate illustrations.